AF504032

LIFE IN 2050

HOW WE INVENT THE FUTURE TODAY

Ulrich Eberl, who was born in Regensburg in 1962, has been interested in the total spectrum of natural science and technology ever since he was a teenager. He studied physics and received his doctorate summa cum laude from the Technical University Munich for a thesis on the first trillionths of a second of photosynthesis. This interdisciplinary work addressed topics in the fields of physics, biology, chemistry, and genetic technology. Starting in 1988 he worked as a freelance science journalist for various newspapers and magazines, writing hundreds of articles on topics ranging from evolution and nanotechnology to space research and the excavation of ancient Troy. After working for Daimler's technology publications from 1992 to 1995, Dr. Eberl joined Siemens in 1996 as the director of its worldwide innovation communications. His particular interest is in futurology, and since 2001 he has been the Editor in Chief and publisher of *Pictures of the Future*, a magazine for research and innovation that has already won several international awards. In a survey of 900 science journalists that was conducted in 2009, Ulrich Eberl was voted the best media spokesperson for corporate research in Germany, Austria, and Switzerland. He lives with his family in Höhenkirchen near Munich.

**FOR MY CHILDREN
THOMAS AND SONJA,
WITH BEST WISHES FOR
THEIR LIVES IN THE YEAR 2050**

www.beltz.de
© 2011 Beltz & Gelberg
in the Beltz · Weinheim Basel publishing group
All rights reserved.
Editorial office: Uwe-Michael Gutzschhahn. English edition editor: Arthur F. Pease
Translations German – English: TransForm GmbH, Cologne
Title: Cornelia Niere; photograph from Archimedes Berlin (www.archimedes-exhibitions.de) /
Oliver Wia („Expedition Zukunft, 2009" science exhibition). Photo: light tunnel of the
Science Express exhibition train of the Max Planck Society.
The tunnel comprises 1,900 LED tiles from Osram
Layout: Irma Schick, Munich
Literary agent: U.M.G. Literaturagentur, Munich
Typesetting: Renate Rist, Lorsch
Litho: ICC-Print, Biblis-Wattenheim
Printing: Beltz Druckpartner, Hemsbach
Binding: Thomas Müntzer printing company, Bad Langensalza
Printed in Germany
ISBN 978-3-407-75357-1
1 2 3 4 5 15 14 13 12 11

TABLE OF CONTENTS

THE FUTURE IN OUR HANDS – THE WORLD IN 2050

It won't be the way it's depicted in science-fiction films. In the year 2050 cars won't be flying through the steel-gray urban canyons of gigantic megacities. We'll still be able to tell robots apart from human beings. Nobody will be beaming astronauts to distant planets. Our brains won't be connected to other brains via radio or telepathy, and no nano-sized repair troops will be zipping through our veins.

Nonetheless, 40 years from now the world will be completely different from the one we know today. Only at first glance do the revolutions that are already shaping up in research laboratories seem less spectacular than the wonders of science fiction. In 2050, tiny sensor and communication elements will be distributed throughout our surroundings; houses and cars will receive sensory organs. Wellness sensors that are almost invisible will measure odors, and smart cameras will warn us about potential accidents. Cars will become mobile robots that find their way almost autonomously, and they will communicate with other vehicles and the electronic mini-offices of their owners. Instead of gasoline or diesel, most vehicles will tank up on electricity and will themselves become power suppliers. They will draw electricity from the power grids and sell electricity to them. Houses will do the same, generating and storing useful energy in many different ways.

The era of large-scale power plants fueled by coal and nuclear energy will come to an end. Instead, there will be hundreds of thousands of installations – be they solar or wind energy installations or geothermal, biomass, or wave power plants – and countless tiny mini power plants providing both electricity and heat in buildings. The era of fossil fuels will be succeeded by a new era of electricity that draws its energy from "green" sources. Cables that stretch across continents and under oceans will create a world-encircling energy network. New technologies that mimic plants will extract the greenhouse gas carbon dioxide from the air and transform it into useful materials for the chemical industry or for biofuels, for example.

The houses of 2050 will be so intelligently built that they will hardly need any additional heat. In their interiors they will have lighted ceilings and transparent walls of light made of glowing plastics, as well as wall-sized displays that open up the three-dimensional world of the new Internet in response to voice commands or gestures. Movies in 3-D will be a matter of course. So will virtual shopping trips, visits to museums, and games played in fantasy worlds – all seeming as real as though you were right there. Universities will offer worldwide learning. Attending a lecture in Tokyo in the morning and a seminar at Harvard the same evening will be no problem, thanks to the Internet of tomorrow.

In 2050, 6.5 billion people of the world's total population of nine billion will live in cities – that's almost as many people as the world's entire population today. Completely new approaches will have to be taken in order to make these cities worth living in. High-rises will become vertical farms; wastewater will be recycled into pure drinking water; T-shirts, packaging, and devices of every kind will be either compostable or designed in such a way that they generate new raw materials instead of waste. Robots will serve as window washers, gardeners, and butlers for senior citizens. Thanks to microchips, blind people will learn to see again and disabled people will regain their ability to walk. Tiny sensors will check blood values in the body and search for migrating cancer cells, so that the doctors of 2050 can intervene promptly if there's a risk of infections, cancer or heart and circulation problems. Many 100-year-olds will then be as fit as the 70-year-olds of today.

All of these things will happen if the researchers of today succeed in transforming the ideas they are currently working on – ideas that will be dealt with in this book – into successful products. If scientists assess trends correctly, if the world escapes major catastrophes, and if people organize their shared existence in a rational way, life in 2050 will not resemble a gloomy science-fiction film. It will be worth living. And here's the best part: We can determine what it will look like. Researchers, engineers, and all the rest of us can help to shape it through the actions we take today.

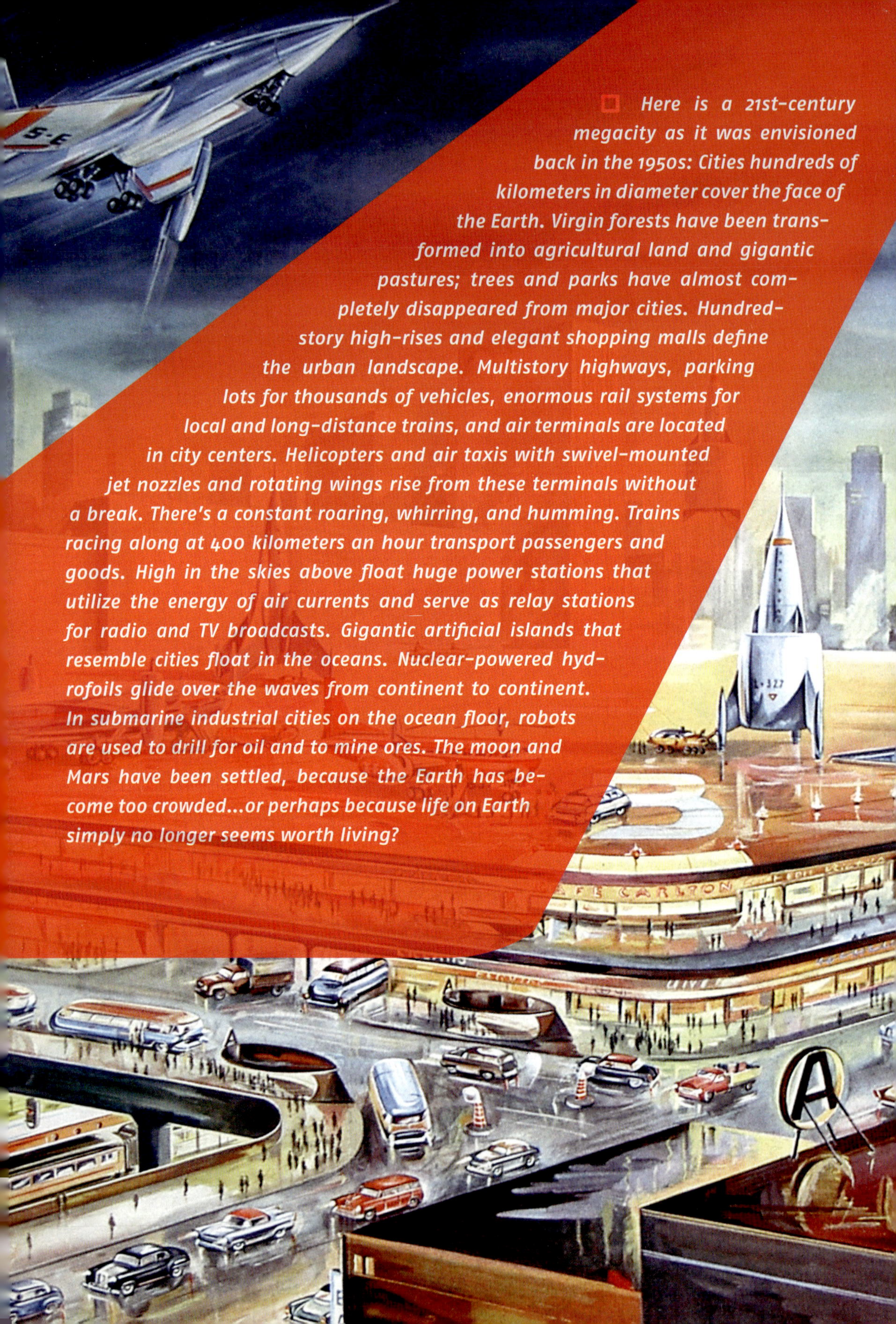

Here is a 21st-century megacity as it was envisioned back in the 1950s: Cities hundreds of kilometers in diameter cover the face of the Earth. Virgin forests have been transformed into agricultural land and gigantic pastures; trees and parks have almost completely disappeared from major cities. Hundred-story high-rises and elegant shopping malls define the urban landscape. Multistory highways, parking lots for thousands of vehicles, enormous rail systems for local and long-distance trains, and air terminals are located in city centers. Helicopters and air taxis with swivel-mounted jet nozzles and rotating wings rise from these terminals without a break. There's a constant roaring, whirring, and humming. Trains racing along at 400 kilometers an hour transport passengers and goods. High in the skies above float huge power stations that utilize the energy of air currents and serve as relay stations for radio and TV broadcasts. Gigantic artificial islands that resemble cities float in the oceans. Nuclear-powered hydrofoils glide over the waves from continent to continent. In submarine industrial cities on the ocean floor, robots are used to drill for oil and to mine ores. The moon and Mars have been settled, because the Earth has become too crowded...or perhaps because life on Earth simply no longer seems worth living?

THE FUTUROLOGIST'S DILEMMA

When an inhabitant of ancient Greece wanted to risk a glimpse of the future, he made a pilgrimage to the navel of the Earth: the Temple of Apollo in Delphi. There the Delphic Oracle, a priestess known as the Pythia, sat on a three-legged stool above a crack in the earth from which steam emerged. In a trancelike state she communicated her prophecies. For example, she predicted to King Croesus that if he crossed the Halys River he would destroy a mighty empire. Croesus, who had always wanted to conquer Persia, obeyed this seemingly good advice — but the empire he destroyed was his own. He had purposely ignored the double meaning of the Pythia's words, as well as the inscriptions carved above the temple's entrance: "Know thyself" and "Nothing too much."

Whether it was the Delphic Oracle, the crystal balls used in the Middle Ages, or the forecasts of Nostradamus — if the prophets made their predictions vague enough, they could not be blamed for false interpretations. Today, however, greater precision is required. Politicians need to know which social, economic or environmental trends they must take into account in their decisions. Company CEOs want to know which products will enable them to conquer the markets of tomorrow. Scientists look for the most promising research fields — and everyone finds it exciting to speculate about life in the year 2050. All of us want something the science-fiction author H. G. Wells called for back in 1900: "a science of the future."

Some aspects of the future are not so hard to predict because many of the strategic directions being set today will have an impact on the world in which we and our children will live in 2050. Many of the power stations that energy suppliers are building today will still be providing heat and electricity. The houses being built today will still be standing several decades hence. The number of children now being born will determine what the age pyramids, and thus the social and health systems, of the year 2050 will look like and how much food, water and raw materials will be needed. And the greenhouse gases we are emitting into the atmosphere today will still have an impact on the Earth's climate in the middle of this century.

But even though we know all these things, there's still room for surprises, as a brief look back will demonstrate. The year 2050 is just as distant from 2010 as 2010 was from 1970. And the futurologists of 1970 made some very startling predictions! The popular juvenile book series Das Neue Universum (The New Universe) teems with gigantic metropolises full of plastic "residence pods", conveyor belts for pedestrians, nuclear-powered hydrofoils, and pneumatic tube systems transporting passengers at speeds of 600 kilometers an hour. The books envisioned industrial cities on the ocean floor even before the year 2000, with "oceanauts" digging for ores. And they predicted enormous settlements on the moon — not to mention the transformation of the jungle, then still regarded by many as a "green hell," into a cornucopia of food for the human race.

A Base on Mars and Captain Kirk's Communicator

Somewhat better grounded than these colorful media visions of the future were the forecasts of the RAND Corporation, a U.S. think tank that still exists today. RAND stands for "Research and Development," and its purpose was to advise politicians and military decision-makers. In the early 1960s, RAND scientists responsible for predicting the future invented the Delphi method, a refined form of which is still being used throughout the world. With the help of questionnaires, experts are asked to evaluate the accuracy of certain theses about future developments and to estimate when these developments will take place. The special feature of this method is the feedback rounds, in which experts are confronted with the assessments made by their colleagues and asked whether they still want to stick to their original opinions. The aim is to arrive at the broadest possible consensus.

In its first Delphi study in 1964, the RAND Corporation predicted that in the 1970s office work would be automated and weather forecasts would be reliable. For the 1980s, the experts anticipated computers that would translate languages, a worldwide satellite communication system, and general acceptance of consciousness-expanding drugs. They expected to see, starting in 1990, raw materials being harvested from the ocean floor and robots being used as household servants and as medical assistants with IQs of over 150. They also anticipated nuclear fusion as a new energy source and a permanent base on the moon.

The study predicted that by 2000 immunization against all forms of bacteria and viruses would be possible and vehicles would drive along highways automatically. For 2010, it prophesied medicines that would increase intelligence and a permanent base on Mars. And finally, for 2020 they forecast the possibility of establishing a direct connection between computers and the human brain, influencing gravity, and slowing down the aging process so as to prolong human life by 50 years. The main risks foreseen by RAND experts were wars waged over energy, food, water, and raw materials, as well as high levels of unemployment and social unrest due to extensive automation.

One interesting aspect of these forecasts is experts' tendency to extrapolate the trends they knew in their own time to fantastic extremes. From the moon landing of 1969 they leaped to a permanent base on the moon and then on Mars; from the first mainframe computer they extrapolated a lin-

*The future as seen in 1958.
A city on the moon with a
plastic dome in the juvenile
book series Das Neue Universum*

guistic genius with the IQ of Albert Einstein; and from the first vaccines they vaulted to the complete eradication of infectious diseases. Another obvious conclusion is that some human fears and visions seem to be timeless. For example, robots that make our work easier but also threaten us are an idea that has haunted us since the 1920s. And the idea of expanding human brain capacity through computer technology is also remarkably resilient. It's still being discussed by some futurologists today.

Nonetheless, some of the RAND researchers' forecasts were entirely accurate, even though they came true later than expected. But it's at least as interesting to note that the predictions made by researchers in the 1960s and 1970s completely missed some developments that characterize the world of today, such as pocket-sized computers and TVs and a worldwide data and mobile telephony network that serves billions of people. The experts at RAND could obviously never in their wildest dreams have guessed that the small communicator used by Captain Kirk, Spock, McCoy, and Scotty in the science-fiction series "Star Trek" could become a reality so soon.

When the World Needed only Five Computers

Another reason why it's so difficult to predict the future is that even experts are not infallible. One example is the now-legendary opinion expressed by Thomas Watson, then the CEO of IBM, in 1943: "I think there is a world market for maybe five computers." Of course what Watson had in mind were the room-sized monsters of his own time. He could not have imagined that only a few decades later his own company would be manufacturing millions of personal computers. Between these two epochs, in 1947, came the invention of the transistor — a tiny silicon component that was to replace the large electron tubes used by mainframe computers.

As late as 1977 Ken Olsen, the founder of mainframe computer manufacturer Digital Equipment Corporation, declared, "There is no reason anyone would want a computer in their home." The very same year saw the market launch of the first industrially produced PC, the Apple II. Even Gottlieb Daimler, the inventor of the combustion-engine automobile, was off the mark when he predicted, "The global demand for cars will never exceed a million, primarily because of a limitation in the number of available chauffeurs." All of these experts were much too bound by their habitual thought patterns. As Albert Einstein once put it, "You cannot solve a problem from the same consciousness that created it."

Belief in the superiority of technological progress was never stronger than it was in the 1960s and early 1970s. People were virtually in love with technology and expected it to guarantee them a life free of care. But it didn't take long for the pendulum to start swinging back. The Six-Day War and the oil crisis of 1973 revealed the precariousness of the world's most important resource, oil. Unemployment rose, and the East-West conflict seemed to be continuously escalating. Nuclear power changed from a panacea – which could even be used to fuel cars – to a threat, and the environmental movement started to take shape.

But many of the negative future scenarios that arose during this period also failed to materialize. Jobs lost to automation were balanced out by new jobs offered by software companies and service providers. For every source of oil that dried up, two others were discovered elsewhere, and the collapse of the Eastern bloc was a development no one had foreseen. Respected futurologists had refused to take into account "such implausible mind games" as the fall of the Berlin Wall. In some cases, the forecasts did not come true because they themselves triggered activities that prevented their fulfillment. For example, the horror scenario of dying forests that prevailed in the 1980s was halted because coal-fired power plants became cleaner, vehicles were equipped with catalytic converters, and foresters employed countermeasures such as mixing acid soils with lime.

In addition to such self-destroying forecasts there were also some self-fulfilling ones. For example, for his film "Minority Report" in 2001, Steven Spielberg asked scientists to speculate about how the world would look in 2054. Thus arose the idea of controlling a computer by means of hand movements. The film's tremendous success apparently inspired developers at Apple and Microsoft to create gesture commands for the iPhone and iPad as well as the multitouch technology of Windows 7. In these systems windows are moved around through finger pressure, and images can be enlarged by spreading the fingers. In this case, the future arrived sooner than expected.

The Major Trends of the 21st Century

All of these examples show how difficult it is to talk about the world of tomorrow with any kind of certainty. Nonetheless, many scientists are convinced that they can at least recognize megatrends correctly. This concept refers not to short-term fashions and hype such as those of pop music, clothes or Internet culture, but to in-depth developments that affect large parts of the world and are practically irreversible. Examples include the globalization of the economy and culture, demographic developments and increasing longevity, urbanization, climate change, and the penetration of every area of life by information and communication technologies.

As early as the 1920s, Russian scientist Nikolai Kondratiev discovered that economic cycles occur in long waves lasting between 40 and 50 years; they begin with important basic innovations that lead to increased prosperity but eventually stagnate and are replaced by the next wave. This concept, which is of course very simplified, vividly describes the phenomena that have shaped the world during the past 200 years. Starting in 1800 these were initially the steam engine and the textile industry. Around 1870 the second Kondratiev cycle reached its crest with the basic innovations of the railroad and the steel industry. From 1900 to 1950, the innovations of electric technology — electric light, streetcars, the radio, the refrigerator, and television — held sway. And the period between 1950 and 1990 was the boom phase of the automobile and the petrochemical industry, oil and plastics.

According to this theory, we are currently at the crest of the fifth cycle, which began around 1990 and is being shaped by information and communication technology, computers, the Internet, and mobile telephony. The sixth Kondratiev cycle, which presumably is slowly starting now, was predicted ten years ago by the researcher Leo Nefiodov. During this cycle, environmental protection issues as well as biological and medical technology are expected to play a vital role. The focus will be on health in the widest possible holistic sense: the health of the environment as well as the health of human beings.

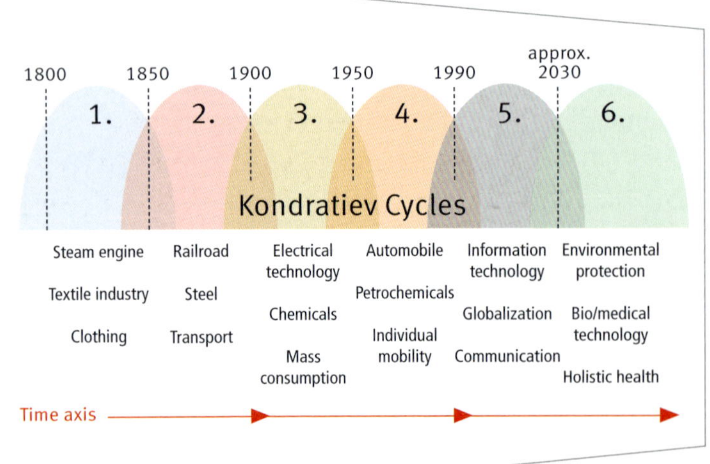

Waves of innovation shape the world. The fifth Kondratiev cycle, today's Information Age, is giving rise to the sixth cycle, which will be shaped by environmental and health issues.

When Futurologists Need Jokers

The megatrends and the Kondratiev cycles provide a rough framework, but futurology becomes especially difficult if we also try to integrate unpredictable leaps — known as discontinuities — into our predictions. The researchers Karlheinz and Angela Steinmüller have developed a method they call "wild cards" or "jokers." In this method, the scientists introduce additional cards into the game, similar to what happens in poker, and investigate the probability of these events actually happening and their possible effects on the economy, society, and politics. The need for such a procedure is shown by events such as the reactor accident in Chernobyl, new illnesses such as AIDS, the fall of the Berlin Wall, the terrorist attacks of September 11, 2001, and the recent worldwide economic crisis. All of them are discontinuities that have undoubtedly changed the course of history.

Crucial wild cards of the future could include major earthquakes and volcanic eruptions, a global epidemic or environmental catastrophe, or the Earth's collision with a comet. Others might be the explosion of a nuclear bomb by terrorists, the use of chemical or biological weapons, a war in the Middle East or one in which Pakistan, India or China is a combatant. Also on the list are unexpected effects of climate change, such as the collapse of major ocean currents, as well as revolutionary inventions such as superconductors that lose no electrical power at room temperature, electrical storage devices with ten times the capacity of today's rechargeable batteries, a breakthrough in cancer treatment, or a way to halt the aging process.

Even though the probability of such wild cards entering the game is slight, futurologists cannot afford to ignore them, because their effects would be enormous. However, even in the case of extremely high-impact events it's worthwhile to take a look at what they have actually changed in the long run. For example, World War II of course brought with it many radical innovations, ranging from the development of the nuclear bomb to the Cold War or the establishment of Israel and the European Community. But how different would the world of today be if Nazi Germany and World War II had never existed? The political structures might be different, but there would be nuclear bombs anyway, because the possibility of their existence is derived directly from the discovery of nuclear fission and Einstein's equation $E = mc^2$. Much of the progress in the development of missiles, satellites, computers, and microchips, as well as in communications and medical technology, would also have taken place anyway, though possibly at a different pace.

Painting Pictures of the Future

In order to take into account foreseeable trends as well as possible discontinuities, about ten years ago engineering giant Siemens AG developed a process known as *Pictures of the Future*. In this process, scientists combine two contrary perspectives: extrapolation from "today's world" and retropolation from "tomorrow's world." The look ahead, extrapolation, corresponds to what companies usually do: mapping the development of the presently known technologies into the future and estimating as precisely as possible at what point in time something will be available. The advantage of this approach — the sure footing provided by starting in the world of today — is also its biggest weakness, because it cannot predict discontinuities. Figuratively speaking, such "road mapping" takes you down a well-constructed highway to the future, but you see little of what's happening elsewhere during the trip — and above all, you can never be sure the highway won't suddenly end, or that you shouldn't have taken another route long ago.

But you can make a better assessment by using a complementary approach, the scenario technique. In this process, you let your mind travel ten, 20, 30 or more years into the future, depending on the focal area. It is certainly much easier to make reliable predictions about the nature of power generation in 20 years than about information and communication technology. For the chosen time horizon, a scenario is designed that takes into account all of the influencing factors: the development of political and social structures, environmental impacts, technological trends, and new customer needs. This makes it possible to think one's way back into the present — via retropolation — and identify tasks that must be accomplished in order to cope with this particular world of tomorrow.

If both of these perspectives are reconciled, we arrive at consistent and internally coherent *Pictures of the Future*. The objective of these processes goes much further than presenting visions and painting a picture of the world of tomorrow. *Pictures of the Future* serve primarily as a systematic process of analyzing the markets of the future, identifying discontinuities, recognizing future customer demands, and identifying technologies that will have a broad impact. Moreover, they also show the course we must take in order to arrive in the future without any detours or accidents. This navigation system guiding us to the world of tomorrow is the key benefit of the *Pictures of the Future* process — it's more a matter of inventing the future for oneself than merely predicting it.

Why We Won't Have Flying Cars

One of the most important conclusions reached by serious futurologists is that in the world of tomorrow not everything that is technically possible will become a reality. Flying cars are one example. They were already part of the future scenarios of 1941, the 1950s and 1960s, and many science-fiction films, and they still exist today, for example in Spaceship Earth at the famous EPCOT Center, the theme park on the future of mankind that was built by Disney in Florida. The many thousands of guests from all over the world who visit the 50-meter-high Spaceship Earth, which resembles a gigantic golf ball, not only discover a great deal about the innovations of the past but also find out about their own future in playful computer-animated video films and see themselves acting in it. And what they see is truly amazing! They eat breakfast in a futuristic energy-saving home, read electronic newspapers, go on vacation in a mini-submarine, are saved after a skiing accident by a robot — and of course travel in flying cars.

The Walt Disney "imagineers" (a concept combining imagination and engineering) who developed these future worlds are a unique team of creative engineers, game developers, and experts from the fields of software, design, and theater. They represent 140 different professional disciplines. When they developed their vision of the future, they were inspired by many details from Siemens' *Pictures of the Future* and added a few of their own — such as flying cars — which come from fantasy worlds rather than actual studies of the future. In other words, they followed Albert Einstein's motto "Imagination is more important than knowledge, because knowledge is limited!" Pam Fisher, a Senior Show Writer at Walt Disney Imagineering, puts it this way: "If we can get the kids who play here to think creatively about their own future, then we've definitely accomplished a lot. The most important thing is to inspire them to make the world a better place, because they are the ones who will one day create the future in which they and their children and grandchildren will live. If you can dream it, you can do it — that's our motto."

Nonetheless, the world of tomorrow will not be teeming with flying cars, even though they have already been technically implemented. For example, the U.S. company Terrafugia builds cars that can fly like planes, and a Dutch company has presented a road vehicle that is also a helicopter. The difficulties lie in completely different areas, ranging from the cost of such a vehicle to legal questions. It would certainly be fantastic to avoid every traffic jam by simply "driving" past it into the air; but this could result in chaos. Special air corridors would have to be defined. And what about safety issues? It would be virtually impossible for every driver to also receive a pilot's license, so the flying would have to be largely automatic — but who would be liable for these flying robots? Another problem would be energy consumption. The fuel burned by cars on the road is being continually reduced, but flying cars would considerably increase fuel consumption, because the entire vehicle would have to be lifted into the air.

In Spaceship Earth at Disney World, there is one overriding theme: the innovations created by the human race yesterday, today, and tomorrow.

People's True Motivations

This is only one example of the fact that many people — especially engineers and natural scientists — prematurely conclude that technical feasibility automatically means quick implementation. But technology is only one aspect of the process by which innovations become widely implemented in society. In addition to issues of cost, safety, and environmental protection, there are all-important social and psychological aspects that accelerate or slow down the introduction of new solutions. Whether or not a technology is accepted depends primarily on whether it supports important human needs, such as the desire for success and recognition or safety and security.

For example, for many people communication with their friends and the respect of their peers is more important than money. This makes it easy to understand the tremendous worldwide success of mobile phones, casting shows, and social networks, apart from any current hype. Whether it's SMS or Twitter, the next top model or pop idol, or communicating via chat rooms, blogs or Facebook, the basic need that is being satisfied is always the same.

These fundamental insights and the megatrends that dominate our century provide the stage on which life will be played in the year 2050. Those who want to see the future in greater depth must take a look inside the laboratories in universities, research institutes, and companies around the globe. Why? Because to a large extent, the world in which we will one day live is being determined by the technologies which are developed in these laboratories today. And that's the purpose of this book: To provide insight into these key areas. What's happening in the fields of energy generation and medical technology? What are the latest developments in information and communication technologies, transportation, and industry? What are the effects of population growth, increasing longevity, worldwide networking, and climate change? In short, how will we live, work, travel, and spend our leisure time in the future? To quote Albert Einstein once again, "I'm more interested in the future than in the past, because the future is where I intend to live."

Morocco, May 2050. Only after the sun has sunk low in the sky does the hawk fly off on its hunt. The heat is still unbearable — the normal temperature is five degrees Celsius higher than it was in the pre-climate change era. The evening brings temperatures of 30 degrees in the shade — but the long rows of parabolic mirrors offer little shade. Still, that's a good thing. After all, the tubes in the middle of the mirrors that focus the sun's rays need to be as hot as possible. The energy from the sun heats up the salt mixture circulating in the tubes to 600 degrees. The heat is sent to a power plant far away on the horizon, where it is used to produce steam. The latter drives powerful turbines that generate electricity. Even during the night, the molten salt mixture contains enough heat to keep the plant operating until the sun rises again. This environmentally-friendly solar facility does the job of five large coal-fired plants. The electricity it generates is used in Morocco to power refrigeration systems, electric cars, and seawater desalination plants that produce drinking water. A large portion of the power is transmitted up north as direct current at a voltage of a million volts. After crossing the Strait of Gibraltar, it travels 2,000 kilometers into central Europe. Energy losses are less than five percent for the desert-produced electricity that is transported to various European cities, many of which are now virtually CO_2-free...

ENERGY FOR THE WORLD: THE DAWN OF A NEW AGE

Things have never been easy for prophets who predict the world is coming to an end. Even back in the days of Troy, no one wanted to believe Cassandra when she warned them about the Greeks and their wooden horse. The consequences were disastrous: Troy, the great city of Cassandra's father, King Priam, was burned to the ground. Cassandra has now returned. For around 40 years, U.S. scientist Dennis L. Meadows has been considered a worthy successor of the ancient doomsayer. In 1972 Meadows published a study called *The Limits to Growth* commissioned by the Club of Rome — an association of economists, scientists, and politicians dedicated to the study of the future. More than 30 million copies of the book were sold. It's one of the most successful books in history — and also one of the most controversial. That's because Meadows and his co-authors predicted the collapse of civilization in the 21st century.

Computer simulations utilized by the authors showed that global population and industrial production would continue to grow largely undisturbed for several decades, but that this would be followed by an abrupt collapse caused by ever-scarcer food and raw material resources and environmental pollution. According to their forecast, this catastrophe would probably occur between 2030 and 2050 — but even under the most optimistic assumptions, it was bound to happen before 2100. Simulations conducted by researchers with updated statistics and figures in the 1990s, and again in 2004, yielded the same results.

A World on the Edge

Only extreme measures to protect the environment, reorganize economic systems, manage birth rates, and recycle raw materials might lead to scenarios that would keep global population and prosperity constant over the long term. However, Meadows isn't very optimistic that the world will be able to agree on such drastic measures: "I expect severe disruptions, much bigger than the financial crisis that began in 2008. I strongly believe that we will see more disruptive changes over the next 25 years than in the past one hundred. My models show tension of the kind seen in earthquake zones. Basically, there's no way of precisely knowing when something's going to happen. What is certain is that there will be a type of quake with severe consequences." Much too much time has been wasted since he issued his first warnings 40 years ago, says Meadows, because politicians and business leaders didn't take his prophesies of doom seriously.

The reason for this is simple. Economic growth is the number one drug of the modern age. Everything seemed to be going well back in the 20th century. Global energy consumption tripled between 1949 and 1972, but oil became less and less expensive as new deposits were continually discovered. It wasn't until the oil crisis of 1973, one year after the initial publication of *The Limits to Growth*, that people began to get worried. "Still, during the 1970s most critics continued to deny there were any limits to growth at all," Meadows explains. "Then in the 1980s they said, 'There may be limits, but they're still far away.' In the 1990s people began to admit that the limits were fast approaching, but they believed there was nothing to worry about because the market would take care of everything."

What they meant was that as resources became scarcer, prices would rise, which is generally correct with regard to oil, natural gas, copper, and steel. Indeed, prices are now rising sharply again even after the economic crisis the world experienced recently. Those who place their faith in the market say that high raw material prices will force the development of alternatives and new solutions such as energy-efficient equipment and cars. That's too short-sighted for Meadows. After all, things can happen so quickly that new solutions arrive too late to prevent a market meltdown.

Among other things, Meadows believes oil production will decline rapidly over the next few decades: "This will happen so suddenly that if we don't act now, we will no longer have any chance of finding alternative energy sources or improving efficiency quickly enough to maintain our standard of living." In fact, the age of oil doesn't even have to end because of oil shortages; it's enough if the risk of exploring for new sources becomes too great. Oil sand mining in Canada, for example, can be just as devastating to the environment as offshore drilling. Remember that it took BP months just to plug a leak 1,500 meters below sea level in the Gulf of Mexico, during which time at least 700 million liters of oil flowed into the ocean. That's the equivalent of 350 Olympic swimming pools. What type of risk will we face when Brazil decides to tap into its huge offshore oil reserves? These deposits are nearly as deep as the Himalayas are high — at least five kilometers beneath a sea that itself is two to three kilometers deep.

The biggest problem with economic calculations is that environmental risk is rarely incorporated into the price of goods. This is just as true for oil production as it is for coal-fired and nuclear power generation. Until recently, for example, it cost a company absolutely nothing to "store" billions of tons of the greenhouse gas carbon dioxide (CO_2) in the air. It wasn't until a few years ago that many countries began experimenting with emission

Environmental disaster 2010: A fire on a BP oil platform illustrates the risks facing companies that are seeking to tap the last of the world's oil reserves.

certificates in a system that requires power plants to pay for every ton of CO_2 they emit above a predefined level. Charging money for carbon dioxide emissions would enable new low-CO_2 technologies to achieve a breakthrough at a much earlier date. However, to function properly, such fee systems would first have to be introduced worldwide in all economic sectors over the next few years. The spectrum of activities affected would cover everything from power plants and industrial facilities to the transport sector and electricity and heat consumption in buildings.

Why We Need Two Earths

According to Meadows, our planet's capacity to supply raw materials and eliminate pollution was already exceeded back in the 1980s. Today humans are acting as if they had 1.3 Earths at their disposal. "If things continue like this, we'll soon need two Earths," he observes. Obviously, even Meadows doesn't know what the future will bring. "I can't tell you what things are going to be like in 25 years," he says. "But I do know that conditions will be tremendously different from those we are familiar with today. Every country in the world is now going through a period of phenomenal transformation."

The fact that the planet has finally woken up from its "more of the same" attitude after many decades, and now takes the modern-day Cassandra seriously, is illustrated by the fact that Meadows is once again a sought-after speaker around the world today. One energy and climate conference seems to follow another, cities and regions are outdoing one another in their efforts to eliminate greenhouse gas emissions, practically all automakers are suddenly looking to replace combustion engines with electric motors, and the solar and wind power sectors are experiencing an unprecedented boom.

This gives cause for hope — and perhaps what Meadows regards as being virtually impossible could in fact be achieved by 2050. It would represent a unique achievement in human history. For the first time, humanity could witness a complete and peaceful reorganization of the basic foundations of an economic system. It would be a new path for nine billion people — one that replaces pillaging with a more careful utilization of natural resources. More than anything else, this also means abandoning oil and coal and establishing an energy sector that produces only minimal amounts of pollutants. Accomplishing such a transformation would require us to shut down thousands of power plants around the world by 2050. These would have to be replaced with hundreds of thousands of wind and solar facilities, and hundreds of millions of electric vehicles would dominate road traffic.

None of this can happen unless people around the globe become convinced that it's the only feasible way to prevent the collapse prophesied by

Meadows and bequeath future generations a world worth living in. This is certainly not an impossible or hopeless task. After all, even though change is sometimes frustratingly slow, people are adopting new mindsets. Environmental and climate protection are buzzwords in Europe, and the principles involved are gaining more and more adherents even in the U.S. and China. What's more, the required technical solutions are either already available or else being developed in research labs, as the following chapters of this book demonstrate. Nevertheless, restructuring will be a Herculean task because we continue to live far beyond our means. Let us therefore take a closer look at how we got here and what actions now need to be taken. To do this, we must first understand what's happening to our climate.

Burning One Million Years per Year

The most severe ravaging of nature ever seen is the work of mankind seeking to quench its thirst for energy. We are annihilating resources that took hundreds of millions of years to form, during which time dead plants and animals sank to the bottom of the shallow seas that covered primordial Earth. This organic material was gradually covered with layers of sediment, leading to an increase in temperature and pressure below it. Completely cut off from the air, organic molecules slowly transformed into hydrocarbons, and ultimately into petroleum and natural gas. Trees, snake grass, and ferns that sank in the moors of the later carbon age also underwent a change that turned them into peat, and later brown and hard coal. Since the 19th century, the human race has removed some 250 billion tons of carbon from the ground in the form of oil, natural gas, and coal — the fossil fuels. Most of this material has been burned to produce energy. Every year, we now consume a volume of fossil raw materials that took between one and two million years to form.

To put it another way, within just a few decades the carbon that plants had bound in organic molecules with the help of sunlight over hundreds of millions of years is being released back into the atmosphere. This is happening in power plants that burn natural gas or coal to generate electricity and heat, in cars that run on gasoline or diesel fuel — both of which are obtained from petroleum — and through processes in which plastic is broken down by microorganisms or burned to produce energy. After all, most plastics are manufactured by the petroleum industry.

It doesn't really matter whether burning or decomposition by microorganisms is involved. Ultimately, the carbon that comes from fossil energy sources binds with the oxygen in the air to create a new compound: carbon dioxide (CO_2). This chemical reaction has two important consequences. On the one hand, it "uses up" the energy contained in carbon or, more accurately, makes it unusable. At the same time, the carbon dioxide that escapes into the atmosphere makes a major contribution to the greenhouse effect that is heating up our planet.

There's nothing negative about the greenhouse effect in and of itself. On the contrary, it is vital to life on Earth because if the process didn't exist, our planet's average temperature would be a very cold minus 18 degrees Celsius. The fact that the Earth's average temperature is much higher — plus 15 degrees Celsius — is mainly due to water vapor, carbon dioxide, methane, and nitrous oxide. These substances act like the glass roof in a greenhouse, allowing sunlight — and in particular its visible shortwave spectrum — to penetrate virtually unimpeded through atmospheric layers and warm the ground and the oceans. This energy is then reemitted as longer-wave heat radiation, but only part of it makes it back into space; the rest is blocked by greenhouse molecules. The blocked heat is reflected by the molecules back to the Earth in what we have now come to call the greenhouse effect — and a large portion of this heat remains trapped here on Earth.

Although scientists know that water vapor is the biggest contributor to the natural greenhouse effect, they still do not fully understand the complex interactions involved. On the one hand, water vapor behaves like a traditional greenhouse gas that absorbs heat rising from the Earth and then returns it to the ground. On the other hand, the clouds that contain water vapor can generate the opposite effect as well. After all, the denser and darker they are, the less sunlight they allow to pass through. Growing quantities of airborne water vapor can trigger a type of intensified reactive feedback: As the Earth heats up, more and more water evaporates out of the oceans and rain forests. If this water reaches the upper layers of air in the atmosphere, it will contribute to the greenhouse effect and cause even more water to vaporize. It's basically a vicious cycle.

As if that weren't enough, scientists have discovered that the sun itself influences the climate in different ways aside from those associated with light and heat. The strengths of the sun's and the Earth's magnetic fields determines how extensively our planet is shielded from the bombardment of the cosmic particles that influence cloud formation. Some researchers even believe that the mechanism at work here can explain short-term temperature fluctuations.

Why the Earth Has a Fever

The natural greenhouse effect is being supplemented by a man-made phenomenon that raises the fever curve of the planet even further. Approximately one eighth of the effect of these additional greenhouse gases is generated by methane, which forms, for example, in the stomachs of cows or when microorganisms decompose organic material. Every cow emits some 200 liters of methane per day when it chews its cud. Laughing gas (nitrous oxide), which accounts for around six percent of the greenhouse gas effect, is primarily caused by intense agricultural activity and excessive use of fertilizers. It is produced, for example, through the conversion of nitrogen fertilizers in soil with a low oxygen content.

However, with a share of more than 80 percent, the largest manmade contributor to the greenhouse effect is the increasing amount of carbon dioxide we are pumping into the atmosphere. Indeed, the combustion of fossil energy carriers and slash-and-burn activity in forests have caused the carbon dioxide content in the air to a rise by more than one third from 280 ppm to 385 ppm since the beginning of the industrial age. Ppm stands for parts per million. In other words, there are now 385 CO_2 particles for every million air particles. That may not sound like many, but it's enough to have caused a 0.7 degree Celsius increase in average global temperatures over the last 60 years or so. Calculations conducted with ice cores in the Antarctic show that concentrations of CO_2 and methane have never been as high in the past 800,000 years or more as they are today.

The World Meteorological Organization has found that the most recent decade was the planet's warmest since systematic temperature measurements began in 1850. The Arctic region is actually nearly three degrees Celsius warmer, which has led to a dramatic reduction of the sea ice. It's possible that the North Pole may soon become completely ice-free in the summer. This actually happened in the Northwest Passage between the Atlantic and the Pacific in 2007. The development raised concerns about the ability of polar bears to survive in the region. The mean temperature in Greenland has risen more than two degrees Celsius over the past 20 years. Every year, 270 billion tons of Greenland's glacier mass melt. That corresponds to a huge block of ice 30 kilometers long, 10 kilometers wide, and one kilometer thick.

Some scientists even believe that global warming will soon intensify, and possibly become more severe, because the buffering effect of the planet's oceans will slowly dissipate. As a result, less and less CO_2 will be dissolved in the sea. In this scenario, the average global temperature could increase by two degrees by 2050 if no countermeasures are taken. Hans-Joachim Schellnhuber, the director of the Potsdam Institute for Climate Impact Research, predicts that our planet will then be warmer than at any other time over the past 20 million years — and this massive change will have taken place within just a century. "That would be a real roller-coaster ride for the Earth," he say. Never before has the climate of the entire planet changed so rapidly. In the past, plants and animals had much more time to adapt to the transition from ice ages to warmer periods and vice versa — in most cases they had centuries or even millennia. Today we might have only a few decades to adjust.

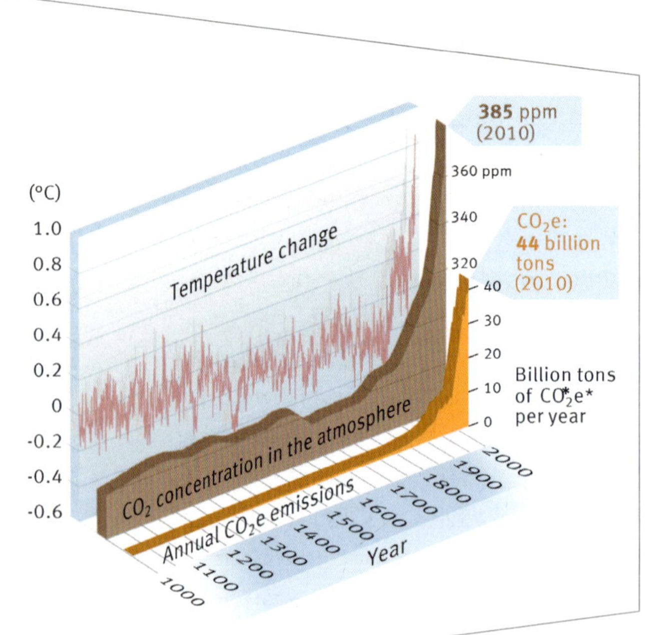

The heart of the climate problem. This famous "hockey stick graph" illustrates that the rise in CO₂ levels and the increase in atmospheric temperature occur in parallel — even though scientists still disagree on the details.

* CO₂e = carbon dioxide equivalent, i.e. all greenhouse gas effects are converted to a carbon dioxide-based scale.

Coastal Cities: Evacuation on the Horizon?

An increase in temperature of just one degree Celsius — something we will probably witness by 2025 — would be enough to significantly increase the incidence of extreme weather phenomena, such as storms, droughts, and floods around the globe. Dry forests could then burn, severe downpours could cause rivers to overflow, mountain glaciers could melt, and many coral reefs and their unique ecosystems could die out due to rising ocean temperatures. A temperature increase of between one and two degrees Celsius — for example, between 2025 and 2050 — would cause a massive encroachment of deserts in Africa and Asia in particular. The oceans would also become acidic due to high levels of CO_2, and the lives of millions would be threatened by crop failure, famine, and a sharp increase in diseases such as malaria. And although the agricultural sector could even benefit in other, cooler regions, being able to produce good wine in Denmark or Norway won't be much help to starving people in Africa in the future.

If no countermeasures are taken, temperatures in the second half of the 21st century could well increase by much more than two degrees — possibly by as much as five to seven degrees by 2100. At that point, we will not be able to manage the consequences. Says Schellnhuber: "In the long term, the sea ice in the Arctic and the Greenland icecap would melt completely, and the Antarctic ice would melt partially. Sea levels would rise enormously as a result. You'd have to evacuate practically all of today's coastal areas, and human civilization as we know it would have to be reinvented." Such a development would directly endanger many major cities such as, for example, Shanghai, New York, Tokyo, Rio de Janeiro, Hong Kong, Hamburg, Stockholm, and practically the entire territory of the Netherlands. Take London, for instance. Here the authorities are worried about what would happen if the sea breached the huge flood barriers on the Thames. Government advisors predict that just one such breach would cause €33 billion worth of damage. Innumerable office buildings, 25 hospitals, and 200 schools would have to be evacuated; large parts of the city's subway, electricity, and telecommunications infrastructure would be crippled; oil tanks would be destroyed; fire storms would occur; and water would become contaminated, spreading disease.

Many experts believe that if we could at least agree on moderate countermeasures, we could keep the global rise in temperature to around four degrees Celsius between now and 2100. That mean value would only translate into a two to three degree increase in water temperature in the Atlantic and Pacific – but it would also cause temperatures on the mainland to rise by five to eight degrees Celsius, and by as much as eight to 14 degrees in northern Siberia and Canada. This would thaw the permafrost soil in these regions, leading to landslides, the formation of giant swamps, and the release of large amounts of methane – in the extreme case, several billion tons of climate-killing gas. That, in turn, would substantially accelerate further temperature rises.

If, in addition, Greenland's glaciers melted completely – a possibility scientists are still arguing about – sea levels could rise by several meters. Hundreds of millions of people would be threatened as a result. The president of the Maldives is in fact already buying up land in India in the belief that his country's 1,200 islands, whose peaks don't even stretch to three meters above sea level, might be lost. Such a population transfer would only be the beginning. In places such as Bangladesh, Laos, Cambodia, Vietnam, and parts of Thailand, it would no longer be enough to simply put houses on stilts and switch over from rice cultivation to fishing, as is already happening today. Instead, entire stretches of land would have to be abandoned.

Other countries in Asia and Africa would also experience a wave of migration, with the problem being exacerbated by refugees from regions plagued by hunger. The rich nations would then have to declare their willingness to absorb millions of these environmental refugees to avert the threat of war. More than one billion people in arid regions would suffer from droughts – not to mention the numerous plant and animal species that would be threatened by extinction as a result of the rapid changes in climate. Some scientists believe that 20 to 50 percent of all species would be significantly endangered by these developments.

Two Tons of Carbon Dioxide per Person per Year

It has now become impossible to completely stop the manmade rise in temperature. Even if we were to put an end to all carbon dioxide emissions immediately, it would take 50 to 60 years for the oceans and forests to absorb so much additional carbon dioxide that temperature would stop rising. A realistic goal, although admittedly a very ambitious one, would be to limit the rise in temperature to a maximum of two degrees Celsius between now and 2100. This would correspond to a CO_2 concentration of around 450 ppm, giving the world a good chance to avoid the most severe consequences of global warming.

The British economist Sir Nicholas Stern has calculated that the measures needed to achieve the two-degree target would cost the equivalent of approximately one percent of current annual global economic output each year. However, if nothing is done, the cost of the resulting damage would amount to five to 20 percent of global economic output — or as much as a 8 trillion per year, the equivalent of the combined output of Germany, France, the UK, Italy, and Spain. The a 400 billion it would cost per year to achieve the two-degree target is also a lot of money, but it's not an unreasonable amount, considering that it would save our world. In fact, it works out to approximately a 33 per person in the world's richest industrial nations — in other words, the European Union countries, the U.S., Canada, Japan, and Australia. These purely economic calculations, which were published in 2006, were a wake-up call for many politicians and industrial leaders. Since that time, the two-degree target has become a global consensus that was written into the Copenhagen climate declaration in 2009 and also adopted as a resolution at the Cancún Climate Change Conference in Mexico in December 2010.

"To achieve the two-degree target, we need to make sure we put no more than 750 billion tons of carbon dioxide into the atmosphere between now and 2050," says Schellnhuber. A look at today's emission levels provides a better idea of what this actually means. Some 33 billion tons of CO_2 are currently released into the atmosphere every year, along with large amounts of methane, nitrous oxide, and other substances that correspond to a further 11 billion tons of CO_2. The total can be broken down as follows: approximately 11 billion tons come from power plant emissions, 5.5 billion tons from the transport sector, five billion from industrial facilities such as steel mills and cement plants, between six and seven billion from slash-and-burn activities and deforestation, and around 3.5 billion through direct emissions from buildings, primarily as a result of heating.

Experts have determined that meeting the two-degree target will require putting a stop to the worldwide increase in greenhouse gas emissions by 2020, and then reducing them significantly afterwards. "We have to succeed in cutting global emissions in half by 2050," says Schellnhuber. Even then, there would only be a 50–50 chance of the temperature increase remaining under two degrees up until 2100. Just how difficult it will be to accomplish that is illustrated by the distribution of global emissions. Every resident of Germany is responsible for around ten tons of CO_2 per year on average, each

American causes emissions of 20 tons — but Chinese per capita emissions are only five tons, and an Indian produces only 1.5 tons. So if every Chinese and Indian citizen tried to attain a Western level of prosperity and use the same amount of energy and emit the same amount of CO_2 as an average German, the result would be an additional 17 billion tons of CO_2 emissions worldwide each year. And if the citizens of China and India became as wasteful as the average American, global CO_2 emissions would double compared with today's levels, and the world would be plunged into a nightmare.

The reality is actually worse, because such calculations only take into account the current number of people on earth. However, there will be more than nine billion people by 2050, given that the global population rises each year by a number equivalent to the population of Germany. If we want to cut greenhouse gas emissions in half, we would therefore have to limit CO_2 emissions to two tons per capita per year after 2050 — an amount that one German emits today simply by driving his or her car 15,000 kilometers per year. Moreover, every American produces two tons of CO_2 a year just by air conditioning and lighting homes and offices. For Germany, achieving the two-ton target would mean reducing per capita CO_2 emissions by around 80 percent by 2050; for the U.S., it would mean a reduction of 90 percent.

The Biggest Challenge of the 21st Century

These figures clearly indicate the kind of revolution that will have to take place over the next 40 years if we are to avoid destroying the foundation of life for billions of people. The greatest challenge in the second half of the 20th century was to end the East–West conflict and avoid a nuclear war. The greatest challenge in the first half of the 21st century will be to find a solution to humanity's huge hunger for energy and resources and eliminate the threat of wars fought for raw materials, water, food, and habitable land.

The fact that mankind was able to master the past century's challenge, which was also considered impossible to overcome at the time, gives cause for optimism. Don't forget that just a few decades ago virtually no one believed the Cold War would ever end peacefully. Even as late as the 1980s, many young people thought a World War III with nuclear weapons and mass annihilation would be the most likely outcome of the Cold War. The slogan "No Future," and the film "The Day After," which depicted the gloomy aftermath of a nuclear exchange, were indicative of the pessimistic attitude of an entire generation.

Yet just a few years later, the Berlin Wall fell and with it the Iron Curtain. Hundreds of thousands of people demonstrated for weeks in the former German Democratic Republic (GDR), ultimately leading to a peaceful transition from one-party rule to democracy. The East Germans took their inspiration from the Poles, Hungarians, and Czechs and the renewal and democratization that swept through the former Soviet Union under the leadership of Mikhail Gorbachev. Just a year after the fall of the Berlin Wall, East and West Germany were reunited, merging at a breathtaking pace that was driven in part by the economic collapse of the GDR and the desire of the German people to live in one nation. None of this would have been possible without the efforts made by prudent politicians in both East and West who were able to avoid an escalation, knowing that rash action could result in disaster at any time.

Today, 20 years later, the Cold War is history and a global nuclear war between the former communist states and their capitalist adversaries is virtually unimaginable. Speaking on November 9, 2009, at a ceremony marking the 20th anniversary of the fall of the Berlin Wall, German Chancellor Angela Merkel had the following to say: "We experienced the fall of the Berlin Wall as an impossibility that suddenly became possible. That gives me reason to say that we should try to do the same in other areas today — for example, to promote peace in the Middle East, in the battle against terrorism, with regard to the international financial system, and in the struggle to halt climate change." The challenges we now face appear to be at least as difficult as those of the last century — but they can be overcome.

Underwater Cabinet Meeting

These days the problem is not how to get politicians from different sides of the divide to agree. The solution to the climate problem — and terrorism and financial crises, for that matter — can only be achieved if all the players agree on the rules of the game. China, for example, now emits more greenhouse gases than the U.S., which is followed by Europe, Russia, India, and Japan in the emission rankings. If you look at Europe not as a whole but instead as individual countries, you'll find that Germany occupies sixth place on the worldwide list of climate sinners. The biggest polluters therefore need to adopt a joint approach and surrender certain decision-making powers to international organizations. If they don't, success will be almost impossible.

Precedents exist — one of them is the World Trade Organization, which brings together more than 150 countries, all of which have agreed to arbitration procedures that in the extreme case can lead to sanctions, with no veto power for any single nation. Nevertheless, there's still too much self-interest and distrust when it comes to the climate issue. On the one hand, there are the dramatic appeals of island nations like the Maldives, whose government cabinet actually held an underwater meeting in diving suits to draw attention to the problem. On the other hand, oil-rich nations such as Saudi Arabia continue to deny that man-made climate change is even occurring.

In between are the U.S. and China, which together account for around 40 percent of greenhouse gas emissions. In order to significantly cut their emissions, the economic systems in these countries would have to undergo dramatic change, something that's already generating tremendous resistance. In addition, China's national pride will not allow it to accept international monitoring of climate protection measures. And then there are the more than 100 developing countries and emerging markets, which rightly point out that they bear practically no responsibility for global warming today. They claim that the highest priority should be given to fighting poverty.

Rajendra Pachauri, Nobel Peace Prize laureate and chairman of the Intergovernmental Panel on Climate Change, says, for example, that it is "unethical and unfair to stipulate CO_2 reduction targets for countries like India, where 400 million people still have no access to electricity." So ultimately, the issue also involves a lot of money that rich nations will have to transfer to poorer ones through a "climate fund," into which the former will need to pay tens of billions each year. The Cancún conference led to the definition of voluntary national CO_2 reduction targets that will be ratified and expanded at the next summit in Durban, South Africa, at the end of 2011. Still, not much time remains for us to act because the consequences of doing nothing will become more expensive and painful with each passing year.

Millions of Eco Jobs

A comparison of the situation today with that of just a few years ago reveals that plenty of progress has been made in the interim. For example, more and more countries and industrial firms are now trying to act in accordance with the principles of sustainability — a term that could become the buzzword of the 21st century. The word was actually coined 250 years ago by German forestry workers to describe a system in which trees were felled at such a rate that the amount of wood removed would not exceed the forest's ability to replace it. The concept was redefined in 1992 during the United Nations Conference on Environment and Development in Rio de Janeiro to mean the following: "Sustainable development meets the needs of the present without compromising the ability of future generations to meet their own needs." Equal weight is given here to satisfying social and economic needs – particularly with regard to eliminating poverty – and conserving natural resources, i.e. environmental protection.

China, for example, not only plans to increase prosperity and eliminate hunger and poverty; the country's government has also decided to invest almost 40 percent of the stimulus program it is implementing to combat the global financial crisis in the establishment of a more environmentally-friendly economy. That amounts to over a 150 billion – almost half of the total investment in green technologies earmarked in all economic stimulus programs worldwide. China has made high energy efficiency a national goal and plans to play a leading role in the development of environmental technologies. Germany, which boasts a 16 percent share of the global market for these technologies, is still number one country in this regard, however. Second place is occupied by the U.S., with a share of 15 percent. Europe as a whole produces almost half of all environmental technology products worldwide.

Industrial companies have long since recognized the opportunities available in this sector and are now competing with one another in an effort to paint their products green and demonstrate that their operations are

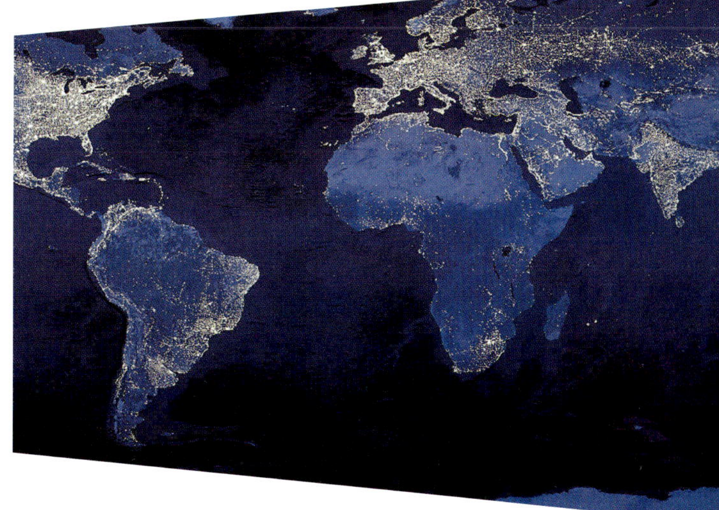

*Bright lights – dark future.
A view from space clearly
shows the places that use the
most energy.*

sustainable. The conventional wisdom today is that environmental technology has the potential to become the most important industry of the 21st century. Companies are already generating annual revenues of more than € 1 trillion worldwide with environmental technologies — and that number could triple in the next ten years. Roland Berger, a business consulting firm, estimates that workforce numbers at German green-tech companies alone could double from one million today to more than two million by 2020. The sector's focus is on renewable energy sources such as wind and solar power, technologies for keeping air and water clean, recycling systems, alternative drive systems, and energy-efficient devices ranging from energy-saving lamps to highly-efficient industrial motors.

Green Industrial Revolution

Governments haven't been idle either. In 2007, for example, the European Union adopted the generally agreed-upon goal of cutting global emissions in half by 2050 as compared to 1990 values. As an initial step, Europe plans to reduce its greenhouse gas emissions by at least 20 percent by 2020 — and will even go as far as 30 percent if other industrialized countries pledge to make comparable reductions. Germany even went a step further and raised its target by an additional 10 percent. The energy concept the German government approved in the fall of 2010 calls for the share of green electricity in the total mix — above all from wind power facilities — to rise to 80 percent by 2050 and the heat requirement in buildings to decline by 80 percent. Power grids and energy storage facilities will also be expanded, and five million electric vehicles are to be put on the road by 2030. Even the U.S., which for decades resisted international climate protection agreements, is now starting to come around. Billions of dollars of venture capital are no longer being invested in Internet technologies in California, for example, but instead in up-and-coming environmental technology. In addition, President Barack Obama is now talking about the need to launch a "green industrial revolution" — even though he still finds it very difficult to take action that would force political institutions in his country to act.

What needs to happen to make such a revolution a reality? Which technology solutions are available for cutting greenhouse gas emissions in half by 2050, despite global population growth and the parallel objective of increasing prosperity? After all, the latter goal requires more energy consumption — which up until now, at least, always meant more greenhouse gas emissions. This is precisely what has to change, however. In the future, it must be possible to increase energy consumption while simultaneously lowering emissions. Experts refer to this as decoupling growth, energy consumption, and CO_2 emissions from one another — for the first time in human history. Basically, there are only three ways to substantially reduce the amount of carbon dioxide released into the atmosphere:

- massive expansion of technologies that generate electricity without emitting CO_2 — this approach includes renewable energy sources such as wind, solar, hydro, geothermal, and biomass power, as well as the development of nuclear fusion power plants;
- the development of solutions that prevent CO_2 from being released into the air, that convert it into other materials, or even remove it from the atmosphere; and
- the most widespread utilization possible of products that use energy much more efficiently — in other words, products that can obtain significantly more useful energy from a kilogram of coal, a liter of oil, or a cubic meter of natural gas.

Engineers and researchers have already made good progress in these areas. The chapters that follow describe many of the exciting associated developments — the things that are already possible and the things that still need to be done if worldwide CO_2 emissions are to be reduced by at least 50 percent between now and 2050.

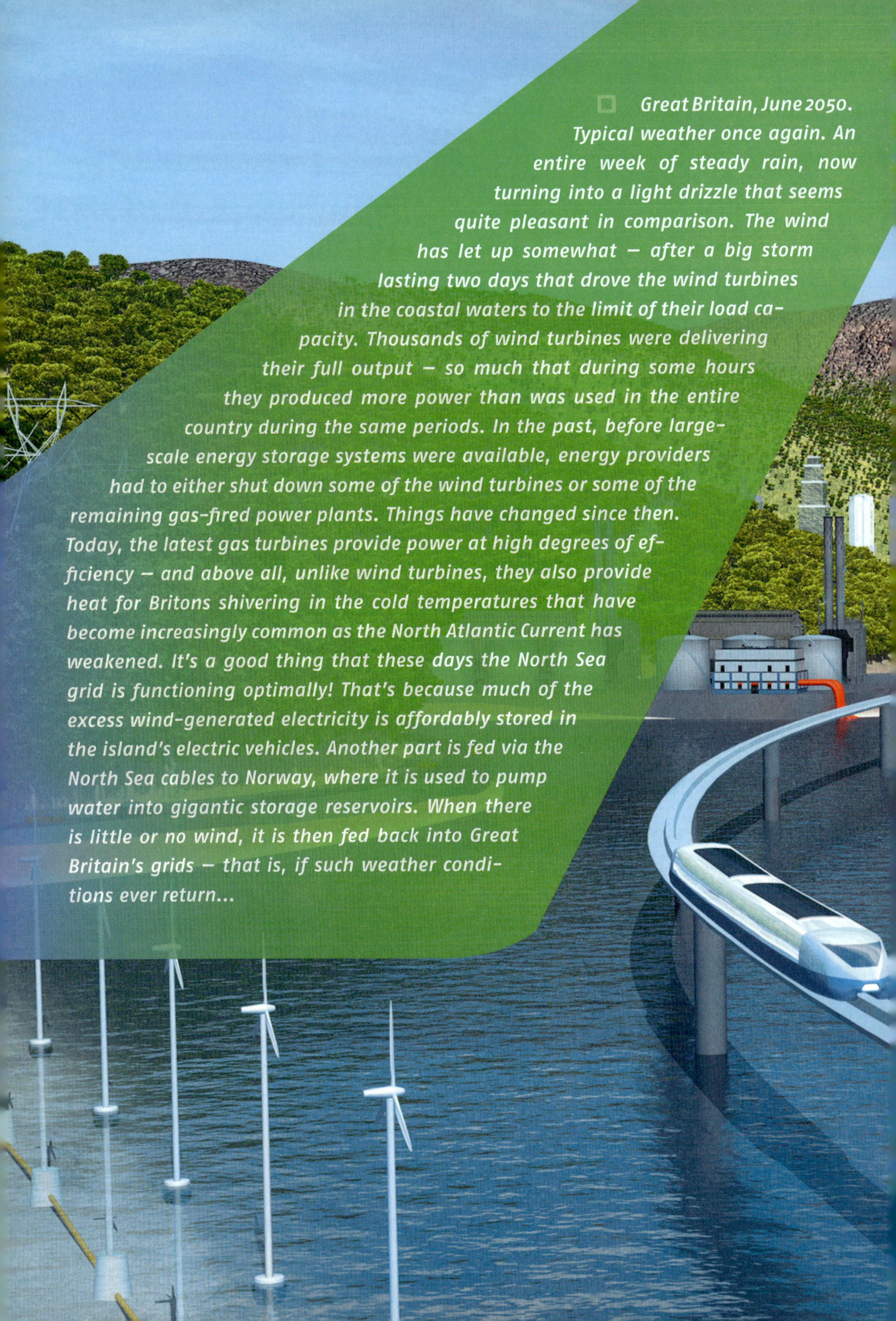

Great Britain, June 2050. Typical weather once again. An entire week of steady rain, now turning into a light drizzle that seems quite pleasant in comparison. The wind has let up somewhat — after a big storm lasting two days that drove the wind turbines in the coastal waters to the limit of their load capacity. Thousands of wind turbines were delivering their full output — so much that during some hours they produced more power than was used in the entire country during the same periods. In the past, before large-scale energy storage systems were available, energy providers had to either shut down some of the wind turbines or some of the remaining gas-fired power plants. Things have changed since then. Today, the latest gas turbines provide power at high degrees of efficiency — and above all, unlike wind turbines, they also provide heat for Britons shivering in the cold temperatures that have become increasingly common as the North Atlantic Current has weakened. It's a good thing that these days the North Sea grid is functioning optimally! That's because much of the excess wind-generated electricity is affordably stored in the island's electric vehicles. Another part is fed via the North Sea cables to Norway, where it is used to pump water into gigantic storage reservoirs. When there is little or no wind, it is then fed back into Great Britain's grids — that is, if such weather conditions ever return...

PIONEERS OF THE GREEN REVOLUTION

Let's visit a few research labs and meet the scientists and engineers who are focusing on developing the energy solutions of the future. All our hopes depend on such professionals, given that they are determined to change the world's energy industry so drastically that by 2050 it will be unrecognizable compared to what it was like in the preceding 200 years.

Henrik Stiesdal is one of these revolutionaries. In the 1970s, while still fresh out of high school, he was greatly concerned about the oil crisis and the discussions initiated by Meadows and the Club of Rome. Stiesdal lives in Denmark, where it's always windy. So why not tap into the power of the wind? Tinkering in his parents' yard, he built one of the world's first wind turbines. He found his materials at junkyards, and the rotor blades were made of wood — "at a cost of fifty cents per kilo," he recalls. In an improved version, his unit then featured three fiberglass blades that automatically oriented themselves to the wind direction. Stiesdal sold the license for his turbine to Vestas, a manufacturer of farming equipment. This transaction made him one of the founders of the global wind boom. Today, Vestas is the world's biggest manufacturer of wind turbines — with about 40,000 wind power plants in more than 60 countries.

But in the 1980s wind power still didn't have a rosy future, and Stiesdal decided to study medicine, biology, and physics. His boyhood dream still had a grip on him, though, and the newly trained biologist was hired by Bonus Energy, a company that was located in the little town of Brande and had metamorphosed from a manufacturer of sprinklers into one of the best-known producers of wind turbines. At that time Denmark was a gold mine of opportunity for wind power pioneers — and it still is today. Even back in the 1980s the government had passed a law regulating energy feeds by wind power facilities into the grid, which led to a boom in the corresponding technologies. "The first machines we built in the early 1980s had an electrical generating capacity of only 22 kilowatts," says Stiesdal. "Capacity has doubled about every four years. Today 3.6 megawatts is typical, more than 100 times as much as that of the first turbines, and there already are five-megawatt machines. And there's no end to this growth in sight."

With almost 100 patents, Chief Technologist Stiesdal powered the rise of Bonus, which was acquired by Siemens in 2004, to innovation leadership in many areas of the wind power industry. He recognized early on, for instance, that wind turbine customers placed particular importance on reliability and the quantity of electricity that can be supplied per year. After all, repairs are very costly and time-consuming — and at offshore wind farms stormy weather often makes repairs impossible for days or weeks at a time. When turbines stand idle and are not supplying power, costs pile up for operators.

Baking Rotor Blades Longer than the Wings of a Jumbo Jet

Stiesdal's main objective is to prevent the need for repairs. "We always have to be able to identify problems before our customers do" is his recipe for success. That's why his team equips wind power facilities with an array of sensors that automatically send the control center not only operating data but also information concerning wind strength, rainfall, lightning strikes, temperatures, and vibrations — often making it possible to determine the risk of outages before they take place. What's more, his team worked for years to develop a wind turbine without a gearbox. This cuts the number of individual parts by half, reduces the weight by 12 metric tons, and makes it less likely that parts will fail and need to be repaired. Another key factor for ensuring excellent reliability is the production process for the rotor blades, which Stiesdal adopted from marine engineering, where many plastic parts are of one-piece construction. Stiesdal, who enjoys sailing in his free time, asked himself why this approach couldn't also be used with wind turbines. The result of his research was lightweight, robust rotor blades that, although they measure 50 to 60 meters — longer than the wings of a jumbo jet — are made entirely without adhesive joining technology or other seams.

Hollow but very strong. Many of today's rotor blades are more than 50 meters long and "baked" in one piece.

The rotor blades are made in a 250-meter-long production hall in shells that look like gigantic sandbox molds. Workers line these forms with thin layers of fiberglass mats and balsa wood. Then they close the upper and lower parts of the forms, after the interior has been filled with inflatable bags to create hollow spaces and make the blades lighter. Next, workers inject several tons of liquid epoxy resin, which makes its way between the air balloons and the fiberglass and evenly joins the two sides of the blade. "And then we bake the whole thing like a gigantic pizza, for eight hours at a temperature of 70 degrees Celsius," says Stiesdal in describing the ingenious production process.

An endurance test conducted by Stiesdal's engineers shows how tough the rotor blades are. An enormous hydraulic arm sets the blades in an oscillating motion, in some cases bending them as much as ten meters up and down. Four million oscillations over a period of just a few weeks simulate the stress resulting from 20 years of actual operation. Such robust blades are necessary because on the open sea, where two to three times more power can be generated than on land, the wind presses air masses weighing

100 metric tons between the blades, which turn in the wind at speeds of up to 300 kilometers per hour at their tips. That's as fast as a high-speed train.

After the rotor blades have been baked and polished, they are transported on articulated trucks to the harbor. There the blades are loaded onto a ship, along with the "nacelles" — to which they will later be mounted and in which the power generation units are found — and the mounting mast. The ship moves everything to its destination — a wind park in the North Sea, for example. Once they arrive at the site everything happens quickly. It takes only six to eight hours to completely install a wind turbine. The construction ship's crane lifts the steel tower, the nacelle, and the rotor onto a pedestal, a steel foundation previously driven 20 meters into the seabed. The components are then bolted together in the good old-fashioned way — by hand.

The North Sea – An Ocean of Wind Turbines

When it comes to offshore wind parks, new records are being set every year. One of the biggest wind parks is currently being installed in the Thames estuary. Plans call for 270 turbines to supply up to one gigawatt of electrical power — enough for 750,000 London households. In January 2010 the British government also announced that it wants to apply investments of a 110 billion to cover one fourth of its electricity requirements by means of offshore wind power facilities by 2020. The roughly 6,000 wind turbines that will be required to meet this target will have an output of 30 gigawatts. Compared to the worldwide electricity mix, almost 70 percent of which still comes from fossil fuel sources, these wind turbines would reduce CO_2 emissions by up to 54 million metric tons a year — about the same amount of CO_2 that is produced by 27 million automobiles.

Today, Denmark already covers over 20 percent of its electricity needs with wind power. And that's not surprising, considering that there are only ten completely wind-free days a year here. On windy days the turbines sometimes produce half of the country's electricity, and on a windy night it's even possible to generate 100 percent. The fluctuating nature of this form of power generation poses entirely new challenges for Denmark. When there are strong winds, for example, the country must either shut down wind power plants or transfer excess wind-generated power to big water storage reservoirs in Norway. When the wind stops, the energy is sent back to Denmark. In the years ahead this storage problem will become much more significant – Denmark, after all, wants to be drawing about half of its electricity from wind power by 2030.

In Germany, wind has been providing about seven percent of the country's power to date, generated by 21,000 wind turbines located mostly along the coast and in hilly areas. In the last decade the number of wind turbines in Germany has more than tripled – enough to make it the world leader until recently, when it was surpassed by the U.S. China occupies the third-place slot, followed by Spain and India. The world's wind power plants have a combined total installed output of 160 gigawatts and can supply almost 300 billion kilowatt-hours (300 terawatt-hours) of electricity per year – as much as 30 large-scale power plants. And the boom is continuing. China alone is planning to generate 100 gigawatts from wind power by 2020, which would make the country the world's biggest market for wind-generated energy.

And there's no shortage of new ideas. Off the coast of Norway, for example, the world's first floating wind turbine is currently being tested. Instead of being installed as a rigid, immovable unit in relatively shallow waters (at depths of no more than 30 meters) as is usually the case with fixed systems, this power plant floats on the surface 220 meters above the sea floor. A 120-meter-long buoy made of steel and concrete and equipped with ballast tanks prevents the wind turbine from swaying back and forth like a bathtub thermometer in rough seas. To ensure that it doesn't drift away, the wind turbine is anchored to the sea floor with three flexible steel cables. The power it generates is sent ashore via a long undersea cable. "This arrangement should be usable in waters up to about 700 meters deep," Stiesdal says.

Power plant on the open sea. Within six hours a construction ship's crane lifts one of the gigantic steel towers, complete with rotors, onto a pedestal.

This makes wind turbine use possible in entirely new offshore locations. Experts have calculated, for example, that more electricity could be generated by wind farms within 50 nautical miles off the U.S. coast than by all of the nation's power plants combined.

Forecasts indicate that the amount of power generated by wind farms worldwide could increase by a factor of thirteen by 2030. However, this would cover just under ten percent of the world's expected energy requirements in 2030, given that electricity use will increase by about 60 percent during the same period. By 2050 wind may be able to satisfy 35 percent of the world's electricity needs. Wind's rapid rate of increase is expected to be surpassed only by solar energy systems, which are expected to increase their collective output 53-fold by 2030, although starting from a much lower level than wind. Today, solar energy generates some 30 terawatt-hours worldwide, but is expected to account for 1,600 terawatt-hours by 2030 — which would equal about two and a half times the total electricity used in Germany today.

A World of Energy in Six Hours of Desert Sun

As of 2010, more than two million solar collectors and photovoltaic systems were operating on rooftops in Germany — the former for heating water, and the latter in the form of solar cells for generating electricity. This boom is due above all to years of government subsidies, which have been used to pay operators of solar cells for every kilowatt-hour fed into the grid — at a rate several times higher than the cost of electricity from coal- or gas-fired power plants. This support has helped to make Germany the world's biggest sales market for photovoltaic systems — ahead of the U.S. and Spain. Photovoltaic systems operating in Germany have a total power output of 18 gigawatts and are currently supplying 11 terawatt-hours of electricity a year. That's enough to cover slightly less than two percent of the country's requirements. In the meantime, however, solar cells, particularly those from China, have become so affordable that in some places solar power can be produced for less than what private consumers pay for electricity from the grid. In the future, this will mean that the power from the rooftops will cost less than that from the grid, even without government subsidies — and more and more homeowners will become their own electricity producers.

So far, however, solar energy has made only a modest contribution to our energy supply. And yet the sun is a true source of power. One ten thousandth of the energy it radiates to earth would suffice to meet humanity's needs. In less than six hours our planet's deserts alone absorb enough solar energy to meet the requirements of the world's population for one year. With this in mind, Gerhard Knies, a retired particle physicist from Hamburg, Germany, envisions harvesting as much solar energy as possible. "It's not only about changing our energy supply because we must do so in order to save the planet — I'm also convinced this holds tremendous business opportunities for companies," he says. Knies, who is 73 years old, has managed in recent years to forge ties with others who are committed to his cause, including experts from the German Aerospace Center (DLR) and the Club of Rome, as well as Prince Hassan of Jordan.

Together, they have developed a concept that Knies named "Desertec." By 2050, according to this plan, power from North Africa and neighboring countries is to not only contribute to the region's energy supply and to seawater desalinization, but also cover between 15 and 20 percent of Europe's electricity requirements. In the summer of 2009 specialists in harvesting power from deserts, achieved what is perhaps a decisive coup. A number of the biggest energy suppliers, solar and electric companies, banks, and insurers founded an industrial initiative for the implementation of the Desertec concept. "The transformation of the world's energy supply can't be performed by individuals with a few solar cells on roofs," says Knies. "Major corporations have to take part."

Mirrors Track the Sun

Desertec won't be inexpensive. Its realization will cost hundreds of billions of Euros. The concept calls for the construction of a number of wind parks and several dozen solar power plants. This won't require a tremendously large area, because, as calculations confirm, the sun provides two to three times more energy in North Africa than in central Europe during the course of a year. In order to generate 700 terawatt-hours a year — or enough to meet 20 percent of Europe's needs — an area of 2,500 square kilometers is sufficient. That's about four times the size of Singapore.

The technology for solar-thermal power plants is simple. Energy from the sun heats water, which vaporizes. This steam is used — as in fossil fuel-fired power plants — to power a turbine, whose motion in a generator creates electricity. The warmth from the sun must be concentrated by means of mirrors in order to achieve high temperatures. The most widely used technology features parabolic mirrors whose curved surfaces point skyward and track the movement of the sun. A receiver tube is fixed in the focal line of the concave mirrors, and a liquid that serves as a heat carrying agent — usually a special oil — flows through the tube. The oil heats up to nearly 400 degrees Celsius and then transfers the heat to water, which becomes steam that drives a turbine. Alternatively, special salts can be used. These salts can be heated up to 550 degrees Celsius, thus increasing the efficiency of the plant. To produce electricity on a sustained basis, future solar power plants will feature heat accumulators that store the captured heat from the sun — for example, in molten salts or in sand — and will thus be able to generate power even at night.

The sun in focus. Spain's Lebrija solar power plant uses 170,000 precisely curved mirrors to supply electricity for 50,000 households.

Solar power plants in the California desert have been supplying more than 200,000 households with electricity for over 20 years. In October 2010 the U.S. government announced that Solar Millennium AG, a German company, would build the world's biggest solar power plant in the Mojave Desert — on an area the size of 4,000 football fields. Plans call for the plant to have an output of 1,000 megawatts, which is enough electricity for up to 750,000 households.

Solar power plants are also being built in Spain, and Morocco intends to be producing several thousand megawatts of power with wind and solar plants by 2020. Electricity from solar thermal power plants costs less than that from solar cells made of silicon, but still significantly more than wind power. Given that large-scale industrial production is just beginning and researchers expect to improve many technical details, in 15 years electricity from solar power plants could be competitive with power generated by conventional facilities. Wind power is already at that stage in some regions.

The New Age of Electricity

Building solar plants in deserts is by no means a new idea. Visionary thinkers such as aerospace engineer Ludwig Bölkow were working on such concepts back in the 1970s and 1980s. However, the focus at that time was on solar cells made of silicon, which were supposed to generate electricity in hot climates and thereby split water, forming the gases hydrogen and oxygen. The plan was to then transport the energy-rich hydrogen via ships or pipelines to Europe, where it would be used in fuel cells to generate electricity and heat, or to use the hydrogen in combustion engines in automotive drive systems. The only "waste product" from the process is water — no carbon dioxide. Hydrogen is such an environmentally-friendly energy carrier that the pioneers of this idea were already elated about a "solar hydrogen era" 30 years ago. But few tangible results can be seen today. Although it's true that companies like Daimler have invested heavily in the development of fuel cell vehicles, and BMW has built powerful hydrogen-fueled combustion engines, hardly anyone is still talking about an hydrogen era.

This is due to technical and organizational hurdles, for one thing, and also to a strong new competitor. Hydrogen is problematic in terms of physics, chemistry, and technology for a simple reason. It is a very volatile gas that is 14 times lighter than air. So in order to use it effectively, the gas must be either greatly compressed or liquefied. But even at a pressure of 200 bar, one liter of hydrogen gas holds only one fifteenth of the energy in a liter of gasoline. In liquid form at minus 253 degrees Celsius it does deliver almost one third of the energy in gasoline, but this requires powerful cooling systems and well-insulated tanks.

A third possibility is to store hydrogen in metal hydrides or "nanotubes," but the storage tanks required for this are not much more convenient. Besides, hydrogen is also a highly combustible gas that can cause powerful explosions if it is exposed to a small spark when mixed in certain ratios with air. Hydrogen is also difficult to handle, store, and transport. It will nevertheless play an important role in our future energy supply, but more probably as a short-term energy storage medium and a chemical intermediate instead of a widely used energy source.

That's a job that will almost certainly be handled by another source — electric current. Its initial success began in the second half of the nineteenth century. In building the first dynamo machine in 1866, Werner von Siemens found the most efficient way to generate electricity. When a coil moves in a magnetic field in such a generator, the mechanical energy — as in a bicycle generator — produces electric current, which can be used for many different purposes, such as lighting or powering an electric motor. Siemens quickly developed many innovations by himself; these included the first electric train, the electric coach — a forerunner of the electric car and the streetcar — and the first electric elevator.

Brilliant inventors on the other side of the Atlantic were also putting the new technology to use. In the late 1870s, for instance, Thomas Alva Edison brought light to homes with the electric light bulb. In 1893 almost 30 million people were amazed by a sea of light consisting of 200,000 light bulbs at the World's Columbian Exposition in Chicago. And two years later, when George Westinghouse and Nikola Tesla opened the first 50-megawatt power plant at Niagara Falls and transmitted current over a distance of almost 40 kilometers to the city of Buffalo, it was the starting signal for the electrification of the world.

Today — 115 years later— the second pioneering age of electrical engineering is beginning, in the opinion of many experts, because electric current is the best way to reduce greenhouse gas emissions. This is because it can be produced with very environmentally-friendly methods, transmitted with high efficiency, and used with minimal losses — making it ideal for the move to a carbon-free energy industry. Unlike the case of hydrogen, the infrastructure for electricity is already in place. And the last missing pieces of the puzzle are now being inserted into the big picture of the "new age of electricity." These include:

- □ new transmission technologies that can transport electricity over thousands of kilometers with low losses,
- □ high-performance energy storage devices for wind and solar energy and for electric vehicles, and
- □ smart power grids that are more flexible and secure than those we have known to date.

Putting the Puzzle Together

Transporting large quantities of hydrogen from the deserts of Africa to central Europe, as envisioned in the Bölkow plan, is very costly. But the Desertec concept is very different. It involves transmitting the energy in the form of electricity via high-voltage direct current power lines, at 800,000 or even one million volts. This is feasible, as has been demonstrated by the first such power lines, which have recently entered service in China to transport electricity from hydroelectric power plants in the country's interior to the megacities on the coast. One of these power lines transmits five gigawatts — equivalent to the output of five large-scale power plants — over a distance of 1,500 kilometers, and the other transmits 6.4 gigawatts over 2,000 kilometers.

Electrical losses associated with these transmission systems amount to only about three percent per 1,000 kilometers. With conventional high-voltage lines based on alternating current, which are used all over the world, the losses would have been two to three times as great. And high-voltage direct current cables have also been laid under the sea, for example from Tasmania to Australia and from the Spanish mainland to the Balearic Islands. In the future, this technology will make it possible to transport power from the Sahara to central Europe — spanning distances of 2,000 to 3,000 kilometers. And this applies worldwide, because 90 percent of the world's human

Giants working to preserve the environment. Transformers as big as one-family houses are needed for the highly efficient transport of power over thousands of kilometers.

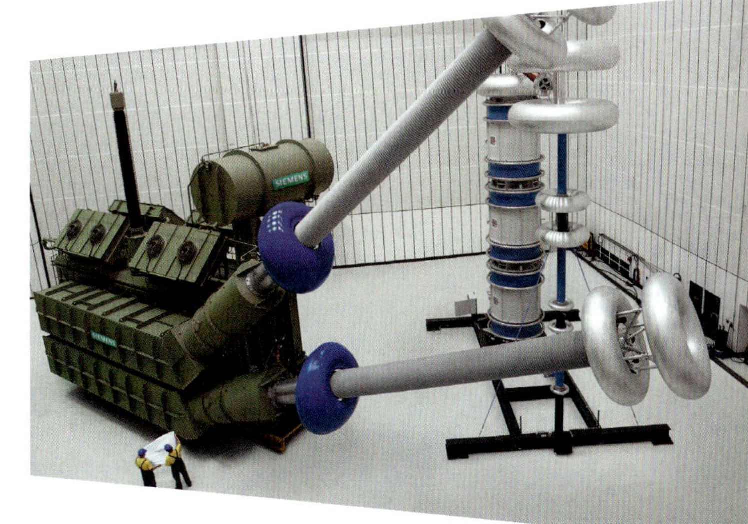

population lives less than 3,000 kilometers from the sunniest regions of the planet, in North and South America, Africa, India, China, or Australia.

Researchers around the world are now filling the second gap in the puzzle of the new age of electricity. Until now, electricity has been difficult to store. This is where the first electric cars failed to measure up. Today, hardly anyone still knows that half of all the motor vehicles in New York around 1900 were electrically powered. Not long before that point, a Belgian engineer named Camille Jenatzy became the first driver to achieve a speed of 100 kilometers per hour — with an electric race car. In 1912, when about 20 manufacturers worldwide built almost 34,000 electric vehicles, the high point was reached. That's when the combustion engine began to catch on. In addition to cheap oil and the invention of the electric starter — which replaced the heavy hand crank — the main reasons for this included the insufficient storage capacity of batteries, which drastically limited vehicle range.

For many decades very little happened when it came to improving batteries — until the tremendous proliferation of laptop computers and mobile phones gave researchers lots of incentive to become active in this field. Today's lithium-ion and lithium-polymer rechargeable batteries can store four to six times as much power, per kilogram of weight, as a classic lead-acid battery. With lithium-ion cells and a storage capacity of 40 kilowatt-hours, today's electric cars must be recharged after about 200 kilometers (see p. 125).

When Energy Suppliers Pay Customers to Draw Electricity

The trend toward increased battery storage capacity is expected to continue. As more and more solar power is fed into the grid in the future, power storage solutions will be indispensable. Today the lack of such solutions is already resulting in very strange phenomena. On October 4, 2009, for example, the price of electricity on the Leipzig energy exchange fell at times to less than zero — to minus a 500 per megawatt-hour. This was due to sudden strong winds that caused the German wind turbines to deliver about 18 gigawatts of power instead of the usual two to eight gigawatts. In short, there was much too much electricity available. And that proved costly for energy suppliers. Not only did each megawatt-hour of power that they fed into the grid generate no income, it even cost the utility companies a 500. They actually should have shut down conventional power plants at that point, but that too involves high costs — and restarting the plants is always a gradual process, which the utilities have to avoid in case the wind dies down just as suddenly as it appeared.

Most of the large-scale power storage systems used today are based on the pumped storage method. Here, water is pumped into an elevated storage reservoir by means of electrical energy at times when there is little demand for power. When demand increases again, the water flows into a reservoir below the elevated one, driving turbines and once again generating power. Today, 33 of these storage facilities with a total output of 6.7 gigawatts are available in Germany. They serve to compensate for peaks in demand — for example in the evening when electric ranges are turned on everywhere. It's almost impossible to build new pumped storage power plants, however, because of a lack of suitable locations.

In January 2010 countries on the North Sea — including Germany, the UK, Denmark, the Netherlands, and Norway — announced plans for a shared eco-electricity grid to help make the storage problem less acute. And there are also plans to invest approximately a 30 billion by 2020 to lay high-voltage direct current cables under the North Sea in order to link wind parks and solar and tidal power plants with one another and with big pumped storage power plants in Norway. Such a network could compensate for weather fluctuations and use the Norwegian hydroelectric power plant as a storage facility, as Denmark is already doing today.

Mobile Power Storage Units — Earning Money with Automobiles

But there is another alternative. In the future it will be possible to use electric automobiles as power storage units. If only 200,000 of them with 40 kilowatts of output each were connected to the grid, the resulting total output would be eight gigawatts — more than the sum provided by all pumped-storage power plants in Germany, and also more than all of Germany requires as a short-term supply cushion to keep the grids stable. This scenario is entirely realistic because automobiles are driven for an average of only between one and two hours a day, and otherwise are stationary in parking lots or garages. What is needed is a sufficiently extensive infrastructure of charging stations and intelligent metering. Vehicle owners could, for example, charge their batteries at night when electricity is inexpensive and then feed it back into the grid during the daytime for peak prices — being sure of course to leave enough power in the battery for the drive home or to stores.

This is what's called a win-win situation. Energy suppliers would benefit because they could use the electric vehicles' rechargeable batteries as power buffers in order to stabilize their grids and thus be able to include more and more wind and solar power in networks. And vehicle owners could earn money with their batteries while also recovering some of the cost of the batteries. This scenario will be investigated during practical testing scheduled to begin in 2011 on the Danish island of Bornholm in the Baltic Sea, where electric cars will draw mainly wind-generated power from the public grid. When demand for electricity rises, for instance at midday or in the evening, vehicles will feed power back into the grid. The Danes hope it will be possible to use tens of thousands of vehicles to compensate for fluctuations in wind power in the future.

Wind in the tank. Electric vehicles that run on power from renewable energy sources are completely CO_2-free.

In addition, work on new power storage solutions is proceeding. Here, materials researchers are urgently needed. Their job is to find new combinations of materials and design batteries that can be manufactured affordably, have high storage capacity, recharge quickly, be environmentally friendly and safe, and have long service lives even under high loads. This combination of challenges is huge, but the incentives are enormous. A market worth billions of Euros beckons to whoever succeeds in developing a power storage system that costs less than € 200 per kilowatt-hour stored – in other words, a 40 kWh unit for under € 8,000 – and who can sell a kilowatt-hour, calculated over the life of the battery, for less than five cents. Today the costs of such batteries are still four to five times higher than this.

Batteries that Run on Rust

A few years ago leading scientists said there was hardly any room to improve rechargeable batteries. But here, as is so often the case, one should never say never. Today, researchers are making exuberant claims for many new ideas. Specialists at IBM Research, for instance, are studying lithium-metal-air batteries that store five to ten times as much energy as lithium-ion batteries. Such batteries, if they can be realized, would make it possible to drive vehicles 800 kilometers before they need to be recharged.

However, there are also safety problems because lithium metal can burn if it comes into contact with water. For lithium-ion batteries, researchers already found solutions. Experts from German company Li-Tec have developed a flexible ceramic membrane that prevents short circuits in a lithium-ion battery in the event of an accident, thus boosting the safety of these high-performance batteries.

Meanwhile, scientists from Korea are using nanotechnology methods to work on microporous silicon, which can absorb considerably more lithium ions than previously used graphite electrodes. And in labs in the U.S., researchers are investigating metal-coated ceramics that are based on the simple reaction of iron with oxygen — in other words, rust formation. They hope to use it to achieve many times the energy density of a lithium-ion cell, and by means of a much simpler production process. So there are innovations everywhere — and it's certainly possible to speak of a boom in batteries.

Another way of storing excess energy from wind and solar plants is hydrogen. Power from these sources could be used to split water into hydrogen and oxygen and to store the compressed hydrogen gas in underground caverns — much like the approach already used today with natural gas. Hydrogen can also be securely stored in these enormous underground cavities, from which less than 0.01 percent a year would be lost. Experts explain that this is because the rock-salt walls of such caverns behave like a liquid, and any leaks seal up automatically.

In Germany there are about 60 underground storage facilities in operation or under construction. If only one of them were used to store hydrogen, it could hold 140 gigawatt-hours of energy, which is more than three times the capacity of all German pumped-storage power plants combined. When stored energy is needed, hydrogen can be turned back into electricity in gas turbines or fuel cells, producing only water as a byproduct — but no carbon dioxide.

Smart Power Grids

In addition to managing the fluctuating supply of wind-generated electricity, power grids will also have to incorporate a growing number of small, regional power producers in the future. Specialists call this decentralized power generation. Here, electricity is produced at many locations by means of solar power units on roofs, newly-developed photovoltaic coatings on walls and windows, mini power plants in cellars, and wind turbines and biomass plants for larger building complexes. This allows more and more electricity to flow not only from large-scale power plants to consumers, as is the case today, but also in the opposite direction, as we will see in the future. Network nodes are created — as in the Internet — that bundle the flow of electricity and steer it in all possible directions. Today's grids aren't ready for this yet, however.

This means that grids have to become smarter. In particular, distribution grids close to end customers — in the 230 to 400 volt range, in other words — are still often a mystery to energy companies. This is because the electric meters used here have not been providing continuous data but are instead read only once a year.

So a lot of information remains inaccessible — data on current energy consumption or the condition of the line system, for example. Up to ten percent of energy disappears from the grid in Europe — either due to inefficiency or electricity thieves — without being noticed by power providers. In the large cities of some developing nations, more than 50 percent of electricity seems to disappear into thin air.

Smart grids — otherwise known as intelligent networks — are the solution. This is the third, and until now missing, piece of the puzzle in the picture of the "new age of electricity," in which powerful computers, sensors, and communication technologies become part of the energy supply system. As a result, the grid becomes more transparent, more flexible, and more effectively controlled. Experts forecast that hundreds of billions of Euros will be invested in smart grids around the world in coming decades. One important element is electronic electric meters, millions of which are being installed today. These meters make it possible for the first time to record in detail where and how much electricity is being used and fed into the grid. Flexible rates then make it possible to adjust use to supply.

A Washing Machine that Knows When to Wash

Until now, the situation has been exactly the opposite. When more electricity is being used, more electricity has to be produced. In the past, many power plants were built only in order to cover such peaks in demand. But in the future, different approaches will be used. Power storage units will step in or energy companies will give their customers incentives to shut appliances off or turn them on later.

When electricity gets expensive and the washing machine, for example, knows this because it features an integrated communication chip, it can then – provided its owner has activated this automatic mode – delay the wash and wait until nighttime to start, when the cost of electricity goes down again. The same applies to cold-storage units or air conditioning systems in big buildings. Here, it makes no difference if a refrigeration unit is interrupted for half an hour and the temperature rises during that time by maybe 0.1 degree.

Another component of the smart grid is "virtual power plants." The basic idea is simple: Small energy producers such as wind and solar plants and hydroelectric, geothermal and biomass power plants, which so far have individually fed their electricity into the grid, can bundle their power and market it jointly. The grid would benefit from this because many small plants, if intelligently controlled, could stabilize the network. As with the Internet, it's not important if one or another component fails or its performance fluctuates – the power finds its way.

In tomorrow's grid, new electricity highways – for example the high-voltage direct current power lines already mentioned – will connect entire continents together, across climate and time zones. This means that by 2050 it will be possible to compensate for fluctuations due to seasonal effects, the time of day, and the weather. It will thus be possible to transmit not only solar power from Africa to Europe or from the deserts of China and North America to these countries' big cities, but also wind power from the high seas in the Atlantic, where the day is just dawning, to the already awake, bustling cities of Central and Eastern Europe.

The trend is clear, though. Electricity will draw the world together. Just as yesterday oil was very much the dominant source of fuel, tomorrow's all-encompassing, environmentally- and climate-friendly energy carrier will be electricity.

Heat from the Earth and Power from Waves

In addition to the wind and sun, a third major CO_2-free type of energy will be in demand: heat from the Earth, which we call geothermal power. It's the only energy source that's available around the clock, all over the world. At any location on the planet, all that's needed is to dig a few kilometers down into the Earth. The rule of thumb here is that the temperature increases by three degrees Celsius for every 100 meters below the ground. Temperatures at a depth of three kilometers range from 80 to 120 degrees; at five kilometers the mercury climbs to between 130 and 160 degrees. This heat can be used in two ways: either directly as hot water from the borehole — which is ideal for volcanic regions like those in Italy, Mexico, and Iceland — or by pumping water into the ground and turning the area there into a continuous flow heater. A geothermal power plant in operation near Munich, for example, provides 6,000 households with electricity and 20,000 with heat. According to various studies, geothermal energy could cover up to ten percent of Germany's energy requirements by 2050, but whether or not this can be achieved will depend above all on how successfully these power plants can supply energy at competitive prices.

The fourth renewable form of energy, hydroelectric power, is already the source of 16 percent of all the electricity generated worldwide. Most of it is produced in power plants at dams and rivers, but significant expansion is no longer possible at these sites— primarily because enormous dams have a big impact on the environment and run into a lot of resistance from the public. However, we have hardly begun to tap ocean energy. Tidal power

Unseen turbines. Energy from tidal currents can be harvested with rotors that work like underwater wind turbines.

plants are used in some places where there are major fluctuations between ebb and flood tides, such as the bay of St. Malo in France, where the tidal range is 12 meters and the turbines in a 750-meter-long dam supply 240 megawatts of power.

By contrast, there's no need for dams with new tidal-flow power plants, which work like an underwater wind park. The density of water is much higher than that of air, so rotor blades don't have to be as big as those of wind turbines, but they do have to be able to hold up under high pressure. The advantage they offer is that tidal cycles make electricity production more predictable. A pilot unit has been in operation since 2008 in the waters off Northern Ireland, where two rotors with a total output of 1.2 megawatts supply about 1,500 households with electricity. Experts estimate that such tidal-flow power plants have a worldwide potential of 200 gigawatts — or one and a half times the installed capacity in Germany today.

While tidal power plants use the power of the moon, which causes the tides' ebb and flow, the wave power plants of the future will be based on the power of the wind. There is a small wave power plant on the coast of Scotland, for example, that sticks to the cliffs like a snail. In a chamber inside the unit, air is compressed and decompressed by the rise and fall of the waves and the pressure difference drives a turbine. This principle is also very suitable for use in harbors, which usually need sea walls to serve as breakwaters anyway. If small air chambers are built into a sea wall, it can also serve as a wave power plant — and at good locations the power thus generated would be about three megawatts of output per 100 meters of wall.

Fuel from Sugar Cane and Electricity from Coconuts

Biomass is the fifth renewable energy source. It provides as much energy worldwide today as wind, sun, water, and geothermal energy combined; but its share of the overall energy mix may decrease in the future. Europe's largest biomass power plant is in Vienna, Austria. It burns wood and supplies 48,000 households with electricity and 12,000 with heat. Burning this fuel releases only the carbon dioxide that the vegetation absorbed from the atmosphere during its growth, which means it is CO_2-neutral — but it is also controversial. The problem isn't the use of scrap wood, as in Vienna, or cow manure as in many developing countries; opponents think it is wrong to grow energy plants on land that could be used to grow food crops or that is cleared by the logging of rain forests — all driven by greed for profits. As a result, food prices then rise and rain forests are destroyed.

This is a complex issue involving a number of factors. Brazil, for instance, is the world's biggest supplier of bioethanol, which is used as automotive fuel and for bioplastics. "In Brazil — far from the rain forest, mind you — four million hectares of land are currently devoted to the cultivation of sugar cane, which yields 22 billion liters of ethanol per year," reports José Goldemberg, a physicist who has played a key role in this achievement. "By 2020 the national market for bioethanol could expand to three times its current size," he estimates, emphasizing that countries including India, South Africa, and Colombia also have suitable climates for growing sugar cane. In India there are also projects designed to use the jatropha plant for biodiesel. The plant's seeds are a rich source of high-quality oil, and it grows in poor soil, even on dry savannas that are unsuitable for any commonly grown food crops.

But the production of biofuel isn't appropriate everywhere. In Europe and the U.S., for example, governments have mandated that it be blended into vehicle fuels. This caused the demand for biodiesel made from rapeseed to rise as much as the demand for bioethanol from corn — which drove corn prices up in Mexico and sparked angry protests. And rightfully so. When it comes down to a choice between using one hectare of land to grow enough food for between ten and 20 people or to produce fuel for one vehicle, the decision should be obvious.

There is much less opposition to second-generation biofuels. These fuels are made from parts of plants that aren't used in food products, but which are well suited for producing energy. Researchers in India, for example, are currently developing small power plants that convert coconut shells into power. They consume about 35 kilograms of this waste product per hour, converting it into hydrogen and carbon monoxide. This gas is converted into electricity in a simple, low-maintenance machine, a Stirling engine with a generator. Such mini power plants can generate between 25 and 300 kilowatts of power, which is entirely sufficient for a typical Indian village with 50 to 100 families. What's more, coconut shells yield coal ash, which can be turned into activated carbon for the preparation of drinking water. This is only one example of a very smart solution that is currently being worked out by scientists in developing countries and emerging economies (see p. 161 f.). In addition to generating electricity and heat from waste products in a CO_2-neutral manner, it is also improving the rural population's drinking water supply and thus preventing diseases.

Both the coconut shell power plant and photovoltaic facilities, whose numbers are increasing in poor countries, are important components of the decentralized power supply of the future. In regions that aren't connected to the big power grids, this is the best way to ensure the availability of electric lighting and electricity for household appliances, televisions, radio, and the Internet. It's definitely much more environmentally friendly than starting up noisy, soot-belching diesel generators. But there is also a trend toward decentralized energy supplies in industrialized nations. In Germany, 100 regions and municipalities with a combined population of more than seven million people have already decided to meet all their energy needs from renewable sources by 2050 at the latest — with wind, solar power, hydroelectric power, biomass, and geothermal power. This will be achieved in areas located throughout the country.

Nuclear Fission or Fusion?

Most experts assume that it will not be possible in the foreseeable future, at least in highly industrialized regions, to do without large-scale power plants that can guarantee sufficient energy to cover basic needs. Such plants can generate energy nearly CO_2-free. For example, some countries continue to rely heavily on nuclear power. In China, India, South Korea, Russia, or wherever — more than 40 nuclear power plants are now in the planning stage or under construction worldwide.

In all, there are 440 nuclear power plants around the world. But in Germany, not one nuclear plant has been built in over 20 years, and plans call for the existing facilities to be gradually decommissioned. Things also seem to be at a standstill in the U.S., where there are more than 100 plants in operation but only one being built. If we look solely at their climate balance, these are exemplary power sources. If the electricity generated by a single large nuclear power plant were produced by typical coal-fired plants, the burden on the atmosphere would increase by between eight and ten million tons of carbon dioxide a year. The difference is clear in one of Germany's neighboring countries. France generates almost as much electricity as Germany does, but its power plants produce 350 million tons less CO_2 emissions per year. That's because almost 75 percent of its electricity comes from nuclear power plants — in Germany the corresponding figure is less than 25 percent.

Nonetheless, by the time the accident at the Chernobyl reactor took place, this type of power generation had already been indelibly stamped with a bad reputation. Even a safety concept with fourfold redundancy, which the latest facilities have, isn't much help against deeply ingrained fears. After all, if an accident happens despite all the precautionary measures, many generations to come would have to live with the effects of radioactivity — and that's not what people mean by "sustainability." In addition, there is still no solution to the problem of how to permanently store radioactive waste. Whether it is put into salt mines or granite caves, no one really knows how safely the waste can be stored at such sites. The time frame here encompasses millennia. The most hazardous element, plutonium 239, which makes up about one percent of spent reactor fuel, has a half-life of 24,000 years — the time it takes for half of the original amount to decay. This combination of the unresolved problem of permanent storage, the questionable safety of the power plants, and the danger of terrorist attacks make nuclear energy an obsolescent model even in the eyes of many energy experts.

Bringing the Sun's Power to Earth

But there may be an alternative. For nearly 60 years, efforts have been under way around the world to produce energy not by means of splitting atomic nuclei as in a nuclear power plant, but rather by joining hydrogen nuclei together to form helium. This process, which is known as nuclear fusion, takes place in the sun's interior. The idea is tantalizing — if nuclear fusion could be used to produce only 500 kilograms of helium per day, the energy released in the process would meet the electricity needs of the entire planet. But the obstacles standing in the way of such an achievement are so enormous that even the researchers who are taking part in the effort say, "Over the years, whenever someone has asked how long it will take until fusion power plants can contribute to the energy supply, the answer has always been '50 years from now.'"

That's what scientists said in the 1970s, when a decision was made to build the first research facility, called JET, in the UK, and the answer still hadn't changed by the end of the 1990s, when JET first produced an output of 16 megawatts — although 24 megawatts still had to be fed into the facility to power the heating. In other words, the plant generated less energy than it used. That situation is expected to change, for the first time, in the ITER international research reactor, which is now being built at the Cadarache research center in southern France and is scheduled to go into operation in the 2020s. "With its 500 megawatts of output, ITER should supply ten times more energy than is needed for its operation," reports Norbert Holtkamp, who as the technical director was responsible for the facility's construction until 2010. If this research reactor is successful, it is to be followed by a large-scale demonstration plant around 2035 — and it might then be possible to build commercial fusion power plants starting in 2050.

One reason it takes so long to build a fusion reactor is technical in nature. Even in the sun's interior, temperatures of millions of degrees Celsius are required before hydrogen nuclei move so close together that they fuse into helium. What's more, the sun's tremendous gravitational force helps to overcome the repulsion of electrically charged nuclei. This gravitational "helping hand" isn't available on Earth, so a fusion reactor here requires temperatures of about 100 million degrees Celsius. Although such extraordinary temperatures are entirely within reach of the right heating system, there are no materials that can contain such a hot gas of electrically charged particles — a so-called plasma.

The sun in miniature. A hydrogen plasma can be heated in ring-shaped facilities until the atomic nuclei fuse, just as they do in the sun.

Bottling a 100-Million-Degrees Plasma

The particles in a fusion plant's plasma are electrically charged, so they can be made to move in a circle in kind of "cage" consisting of a magnetic field that keeps them away from the walls. To accomplish this, engineers need very strong, specially-shaped magnets with superconducting coils, which have to be cooled to temperatures of minus 260 degrees Celsius. The special materials, the superconductors, ensure that current flows at these low temperatures without electrical resistance and can produce magnetic fields with very high density. This magnetic "bottle" must be so ingeniously designed that the plasma's internal temperature of 100 million degrees falls to a relatively moderate 250 degrees Celsius at the walls within a space of just a few centimeters.

A fusion power plant of the future would thus have an onion-like design, with several walls and a vacuum vessel, around which the magnetic coils are wound. The fuel, consisting of the hydrogen nuclei deuterium and tritium, would be injected into the plasma in the form of frozen pellets. A large power plant generating, for instance, three gigawatts of electrical power, would consume about 100 grams of the fuel per hour. Such a power plant would require heating for a few seconds during its start-up phase, until the deuterium and tritium fuse and the fast helium nuclei — caught in the magnetic field — begin to race around. The nuclei would transfer their energy to the plasma by means of collisions and thus continue to raise its temperature.

From that point the fusion reaction would burn on its own. The neutrons produced during nuclear fusion are electrically neutral — they have no charge — so the magnetic field can not prevent them from leaving the plasma. They therefore shoot unimpeded into the walls and heat them up, which is the purpose of the power plant. In the walls, a conventional cooling agent, such as water, absorbs the heat, turns it into steam, and generates electricity in traditional turbines and generators.

The entire process produces no carbon dioxide, and radioactive waste is minimal. It is primarily the walls of such a reactor that would require interim storage following decommissioning. However, their level of radioactivity would diminish very quickly. After 100 years it would decline to only one ten-thousandth of its original level. Fusion reactors would also be very safe. "If the plasma comes into contact with the wall, its temperature drops immediately and the plasma reaction stops. This is because the reaction requires high temperature to power the fusion," Holtkamp explains. "It's possible that part of the wall could be destroyed, but definitely not the reactor itself. There wouldn't be enough energy in the plasma to do that." Because of the physics involved, an accident with catastrophic consequences that could endanger the lives of people living near a fusion reactor would be impossible.

The fuel accounts for a negligible share of the costs of a fusion power plant. Deuterium is extracted from water, and tritium is produced in the reactor, from lithium, of which the world has a plentiful supply. The bulk of ITER's currently-estimated cost of a 12 to a 15 billion is earmarked for complex power plant components and the facility's construction. Much of the expense is due to the fact that all the participants must agree on how to move forward. That's because the ITER research project is being developed, built, and operated by seven partners who have equal decision-making power but are very different. They include the European Atomic Energy Community, Japan, Russia, China, South Korea, India, and the U.S. Given this situation, it's no surprise that the project was named ITER — which not only stands for "International Thermonuclear Experimental Reactor" but also means "the way" in Latin.

Building a Coal-Fired Power Plant Every Two Days

Considering all these difficulties, it is unlikely that fusion reactors will be covering a significant share of the world's energy needs by 2050. A more likely scenario is that they will be making their first contribution in the second half of the 21st century. On the other hand, another class of large-scale power plant — coal-fired and gas-fired plants — will surely still be playing an important role in 2050 for the simple reason that they currently provide the lion's share of our energy and will most likely remain in service for several decades.

Today, these plants produce almost two thirds of all the electricity generated worldwide. And the International Energy Agency (IEA) estimates that they will still be contributing more than half of the world's power supply in 2030, despite the boom in energy from renewable sources. The IEA isn't offering any in-depth forecasts for the period beyond 2030 — only various scenarios. But one factor can't be dismissed: If we consider the long service life of such power plants and the available reserves of coal and gas deposits — which could keep prices relatively low — it's a good bet that in 2050 more than one third of all the power generated worldwide will still come from coal and gas.

However, electricity requirements could easily double by 2050, due to population growth and increasing prosperity in emerging economies, as well as the expected boom in electric automobiles. This leaves us with a clear conclusion. Today, 62 percent of the total 20,300 terawatt-hours (TWh) of electricity produced annually worldwide comes from coal-fired and gas-fired power plants, which amounts to about 12,600 TWh. If 42,000 TWh are needed by 2050, with one third of it coming from coal and gas, that equals roughly 14,000 TWh. In absolute numbers, this amount is more than we are consuming today, even though the relative share of coal and gas in the power generation mix will be half of today's level, thanks to solar power, wind power, hydroelectricity, biomass, and geothermal energy.

And this is an optimistic prognosis, considering that the boom in coal and gas power plants currently shows no signs of diminishing. In 2006, China put 174 coal-fired power plants into service — an average of one plant every two days! Putting this in perspective, in one year China expanded its energy supply by more coal power plant capacity than Germany has in total. China's plant construction has slowed in the years since then, but the country is still generating 73 percent of its power with coal. In Germany, coal accounts for 43 percent of the mix.

The reason for this is obvious. China has an ample supply of low-cost domestic coal. It extracts 2.5 billion tons a year, almost 45 percent of all the hard coal mined worldwide — and these numbers are rising rapidly. China's CO_2 emissions are keeping pace with this increase. The Chinese government likes to announce that the country is responsible for less than ten percent of the carbon dioxide that has been put into the atmosphere by humans — while the U.S. is responsible for more than a quarter of the emissions — but if the current trend continues, China will be producing a major share of the additional emissions in coming decades.

A key step in counteracting such a development is China's decision to increase its share of energy from renewable sources — wind, solar power, biomass, and especially hydroelectric power — from seven percent today to at least 15 percent by 2020. But sharply cutting the CO_2 emissions of its coal and gas power plants will be even more vital. There are two ways to achieve this: either make the power plants more efficient — in other words generate just as much power while using less coal and gas — or prevent the carbon dioxide that is released by combustion from getting into the atmosphere.

Rigorous Testing of Materials

A lot of progress can still be made by means of the first measure — boosting efficiency. The best of today's coal-fired plants are so good that they deliver one and a half times as much electricity as the worldwide average output, while using the same amount of raw materials. China, with the help of Western companies, recently put such power plants into operation near Shanghai. These plants have a 45 percent level of efficiency, which means they convert 45 percent of the energy in coal into electricity. The worldwide average efficiency is a little over 30 percent. The new Chinese power plants are therefore better than almost any of existing facilities in Europe, where the average efficiency level of coal-fired plants is no more than 38 percent.

Every percentage point gained with a large 1,000-megawatt power plant represents a savings of 180,000 tons of CO_2 a year. As a result, the 15 percentage points by which one of the new power plants surpasses the international average represents 2.7 million tons less CO_2 in the atmosphere. This is made possible primarily by new steam turbines, which are driven by the hot steam produced by the combustion of the coal. The turbines run at 600 degrees Celsius and a pressure of over 260 bar, which enables highly efficient power generation — the higher the temperature and pressure, the higher the efficiency.

Using new materials and manufacturing processes, engineers intend to achieve temperatures as high as 700 degrees and a pressure of 350 bar between now and 2020 — which should result in an efficiency of 50 percent. However, the stresses this exerts on the materials — the equivalent of having to endure temperatures of 700 degrees Celsius on the sea floor at a depth of 3,500 meters — are far too great for conventional metals.

Researchers are instead relying on sophisticated alloys of high-strength metals including nickel and chromium, with just a "pinch" of iron. But this requires new casting, forging, and milling processes that will first have to be developed for the parts, which weigh tons. Last but not least, it must be proven that they can hold up under the enormous stresses, because customers demand a steam turbine that will run for at least 200,000 hours — a service life of more than 25 years.

Just One Turbine for 1.8 Million People

Nickel alloys have been in use in gas turbines for quite some time. But compared to gigantic steam turbines, gas turbines running on natural gas could almost be described as delicate. While their interior temperatures of up to 1,500 degrees Celsius are considerably higher, the pressure, at 20 bar, is relatively low. Very high temperatures significantly boost the efficiency that can be reached with such power plants. If natural gas is burned in such a gas turbine for power generation, and the exhaust gases, which are still at 600 degrees, are then used to turn water into steam and generate electricity once again with a downstream steam turbine, an efficiency of over 60 percent can be achieved in this type of combined-cycle power plant. If the heat thus created is used for district heating of buildings or to supply heat to industrial facilities, the result is an overall efficiency of about 90 percent. Power can hardly be generated more efficiently than that.

The world's most powerful gas turbine is 13 meters long, five meters high and tips the scales at 444 tons — that's almost as heavy as six diesel locomotives. Siemens put this colossus into operation for the first time in late 2007 in a new E.ON power plant in Irsching, near the city of Ingolstadt in Bavaria. This gas turbine's 375 megawatts of power equals the output of 100 large wind turbines. It could supply enough electricity for a city of 1.8 million people — more people than live in Vienna or Philadelphia, for instance. Together with a downstream steam turbine, the plant reaches an output of 570 megawatts.

With its efficiency of over 60 percent, such a power plant emits almost 75 percent less CO_2 per kilowatt-hour of electricity than average coal-fired facilities. Such natural gas power plants will therefore have their place in a green future — especially since a gas turbine can be started up within five minutes from standby mode and deliver full output after just fifteen minutes. This makes them ideally suited for offsetting the natural fluctuations in wind and solar energy.

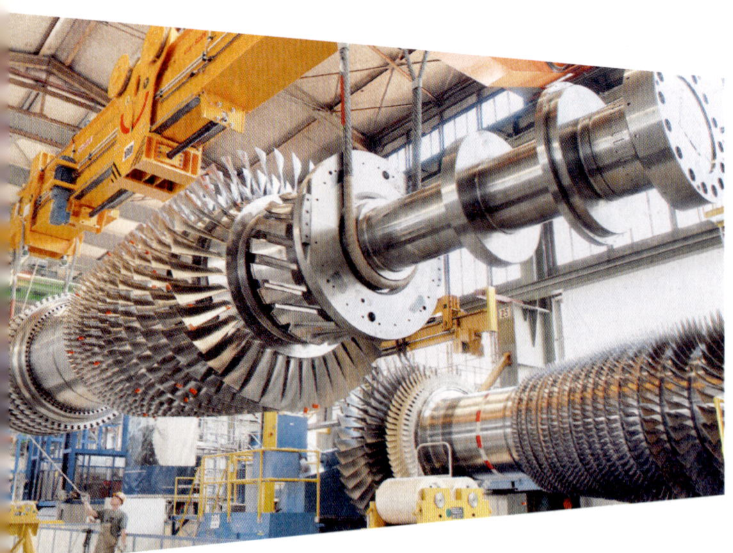

As heavy as six diesel locomotives. The world's biggest gas turbine has an output equal to that of 100 large wind turbines.

More than 800 developers worked for nearly ten years on the construction of the Irsching turbine, which consists of over 7,000 individual components. At a gas turbine plant in Berlin, tiny elements and parts weighing several tons were assembled with a watchmaker's precision and optimized down to tolerance measurements of thousandths of a millimeter. "It's tricky when you want to send a tornado of gas heated to between 1,200 and 1,500 degrees Celsius over metal turbine blades," says Willibald Fischer, the head of the development team. "That's because the highest temperature the blade surfaces are allowed to be exposed to is only 950 degrees. At that point the surface begins to glow red. If it gets any hotter, the material begins to lose its stability and oxidizes."

The turbine's engineers came up with some creative solutions to prevent such a situation. The turbine blades are hollow and in the hottest part of the turbine the blades have fine holes from which an insulating film of air flows over the blades and covers them. In addition to this active air cooling, they also have two thin protective layers: a 0.3-millimeter-thick undercoating directly on the metal and a thin ceramic layer on top of that, which provides the actual heat insulation. The undercoating requires an especially clever design. It serves as an intermediary between the ceramic and the metal and also prevents oxygen from the combustion air from coming into contact with the metal and allowing it to rust — or "oxidize," as the experts say.

A Pinch of Rhenium

Werner Stamm, a materials researcher in Mülheim an der Ruhr, Germany, is a true artist when it comes to creating the formulas for such materials. His latest discovery for the optimal undercoating is rhenium, a rare metal with a very high melting point of 3,200 degrees Celsius. In small quantities of from one to two percent, it improves not only the mechanical properties of Stamm's previous mixture of cobalt, nickel, chromium, aluminum, and yttrium; it also prevents the aluminum from gradually penetrating into the base material. "Without this protective coating the nickel alloy would survive for only 4,000 hours," says Stamm. "With the coating, however, the alloy can hold out against the oxygen in the combustion gas for more than 25,000 hours. That's longer than the time demanded by power plant operators for such turbine blades."

But it isn't only the temperatures that put the blades under stress. Centrifugal force is also tremendous. The tip of each turbine blade is exposed to forces of up to 10,000 times the acceleration due to the Earth's gravity, which is the same as if each cubic centimeter weighed as much as a fully grown adult. Traditional manufacturing processes are outdated when it comes to making these high-performance components, so special cooling processes are used to make them very strong and resistant to breaking. Engineers also use computers to optimize the shape of the blades with the help of 3-D simulation programs. Each of these measures results in only a fractional improvement in efficiency, but all of them combined make new world records possible.

In tests conducted over a period of 18 months, the Irsching turbine was equipped with 3,000 sensors that measure just about everything modern technology can register today — from temperature and pressure to mechanical tension and material strain. Normal operation is scheduled to begin in 2011, and the turbine will then be offered to power plant operators around the world. The high-efficiency turbine helps to not only protect the environment but also to realize cost savings, because it needs much less natural gas per kilowatt-hour generated. An energy supplier in the U.S. that has already ordered six of these units is hoping to save about $1 billion in operating, maintenance, and investment costs.

In the coming decades, researchers like Werner Stamm will be trying, among other things, to make even hotter temperatures possible and to achieve even higher efficiencies. The ultimate dream of such researchers is blades that are not made of metal, but are instead entirely made of ceramic material. These could dispense with cooling completely, because ceramics can withstand much higher temperatures than metals. But that's still a long way off, says Stamm: "Maybe in 15 years — but people in this field were also saying that 15 years ago."

This is mainly because all ceramics have a drawback, as everyone who has ever dropped a teacup or a plate on the floor knows. They're very brittle and break easily. That's why they need to be reinforced if they are to hold up for the required 25,000 hours. The most promising current idea is to use fiber-reinforced ceramics. Glass-like fibers made of aluminum oxide and silicon dioxide keep a ceramic material intact, even if it has cracks. However, such fibers reduce resistance to high temperatures, as they can withstand only 1,200 degrees — whereas the ceramic alone would be able to take up to 1,700 degrees Celsius.

How to Cut CO$_2$ Emissions by as much as the European Union Produces

Concepts like these can make coal-fired and gas-fired power plants even more efficient. But tremendous progress has already been achieved to date. Per kilowatt-hour of electricity, for example, a coal power plant with the future 700-degree technology would use only 288 grams of coal and emit 669 grams of CO$_2$. Today's coal-fired plants, by contrast, need an average of 480 grams of coal and produce 1,115 grams of CO$_2$. Combined-cycle power plants are even better. A facility equipped with the new gas turbine would emit only a little more than 330 grams of CO$_2$ per kilowatt-hour. The following calculation clearly shows the potential that these improvements hold: If all the world's coal-fired power plants today had the 700-degree technology, their CO$_2$ emissions would be 3.7 billion tons less than the current level. An additional 300 million tons of CO$_2$ emissions could be prevented if all gas-fired power plants had turbines like those in Irsching. This potential reduction of four billion tons roughly corresponds to the entire CO$_2$ emissions of the European Union!

The U.S. West Coast, July 2050. Buying emissions rights has become very expensive — 50 dollars per metric ton of CO_2. At that rate, one of the large, old coal-fired power plants would have to pay 400 million dollars every year just to emit the greenhouse gases it normally generates during operation. But there aren't many of those carbon spewers left. Today, you see solar power plants and wind turbines everywhere, even high up in the spaces between skyscrapers. And why not? After all, it's pretty windy up there. And the large-scale power plants now operate equipment that removes the CO_2 from their exhaust gases. Although that requires quite a lot of energy, it is worth the cost, considering the prices that power companies would otherwise have to pay for emissions rights. Some of these plants are pumping the CO_2 into petroleum deposits under high pressure in order to extract the last remains of that precious raw material. Others are using it in huge algae tanks to accelerate the growth of the tiny green organisms from which they can make biofuels and bioplastics. There are also a few companies that specialize in removing CO_2 from the air directly, either by generating electricity from wood chips and organic waste and capturing the resulting CO2, or by using the electricity generated from wind energy to produce hydrogen. The hydrogen can either be used directly or converted to methane with the CO_2 of the air, and both options are ideal ways of storing energy for a CO_2-neutral world...

DON'T LET GREENHOUSE GASES ESCAPE!

Engineers and researchers are very inventive when it comes to squeezing the greatest amount of electricity out of a ton of coal or a cubic meter of natural gas. But all of this helps the climate little if, at the same time, hundreds of new coal-fired and natural gas power plants are being built — particularly if the additional CO_2 emissions are greater than the cuts achieved at other locations. In this case, there are only two solutions: Either the CO_2 must be captured before it reaches the air, or it must be removed from the atmosphere afterward. Just looking at a garden should be enough to convince us that this is indeed possible. All green plants grow by extracting CO_2 from the air and incorporating the carbon into their biomass. This has inspired scientists around the world to dream of creating a "green" power plant that burns coal or natural gas. They want to separate the CO_2 generated by power plants and either store it safely or convert it into useful materials by using it to feed algae, for example.

The separation of CO_2 produced in coal-fired power plants is already being tested in pilot installations. In one variant, the coal is converted into a combustible synthesis gas from which the CO_2 can be removed with relative ease. The fuel that is ultimately burned is then pure hydrogen, and only water vapor is emitted into the atmosphere, no CO_2. In other test systems, researchers scrub the CO_2 out of the flue gas of the power plants after combustion takes place. They channel the emissions through a large tower containing a scrubbing solution through which the gas bubbles upward. This solution contains special salts that combine with CO_2 very effectively. In fact, 90 percent of the greenhouse gas can be captured this way. Finally, in a different part of the system, the CO_2 is removed from the salt through heating and then compressed. Afterward, almost all of the salt can be reused and fed back into the cycle.

All of these methods of CO_2 removal are elaborate, cost a great deal of money and energy, and currently lower the efficiency of power plants by about nine to 12 percentage points. In other words, they do remove the greenhouse gas, but they simultaneously nullify the efficiency gains achieved over the last 20 years. So far, in other words, it hasn't been technical hurdles that have prevented large power plants from using CO_2-separation technologies. Without government subsidies, as in the case of the pilot projects, it simply isn't worth the cost at present.

The economics of CO_2-separation technologies for power plant operators will change only when it becomes more expensive to emit CO_2 into the atmosphere than it is to remove it before or after the fuel is burned. Such a development depends on the one hand on how expensive it is to build CO_2-separation systems and on the other hand on the price of carbon credits for a ton of CO_2. Power plant operators are required to purchase these emission rights if they intend to release greenhouse gases into the atmosphere. Such rights currently cost between 10 and 20 euros per metric ton of CO_2, which is much too cheap — experts estimate they will probably rise to over 30 euros between 2020 and 2030. It would then likely be cheaper to remove the greenhouse gas than to blow it into the air.

Storage a Thousand Meters below the North Sea

In addition to the economic hurdles associated with CO_2-separation, however, there is another obstacle to overcome: the political issue. It is still altogether unclear what should be done with the huge quantities of CO_2. Currently, the focus of most discussion is subterranean storage. There is said to be plenty of room deep in the Earth. In Germany alone, experts estimate the capacity for CO_2 storage to be 12 to 28 billion metric tons. That would meet the needs of German power plants for several decades. The United Nations

Intergovernmental Panel on Climate Change calculates that the world's former petroleum and natural gas deposits could hold 900 billion metric tons. And ten times that much CO_2 could be stored in saline aquifers — layers of sandstone saturated with salt water. In theory, these aquifers have enough capacity to store all of the CO_2 that human beings could ever release from burning fossil fuels!

The best solution would be to use pipelines to transport the gas directly to these storage sites from the power plants where it would be separated and compressed or liquefied. This is feasible, as shown by the Norwegian oil and gas company Statoil Hydro, which has pumped ten million metric tons of CO_2 into a stratum of rock 1,000 meters below the North Sea since 1996. The CO_2 is extracted along with the natural gas as an undesired constituent. It would be expensive for Statoil Hydro to simply release CO_2 into the atmosphere, because Norway imposes a high tax on every ton of the gas emitted in this way. Researchers have been monitoring this storage of CO_2 underneath the North Sea, and the results are clear. The CO_2 remains stable in the reservoir and does not escape upward in the form of gas bubbles, for instance.

In Germany, a similar project has been under way since 2008 at Ketzin, near Potsdam. There, 60,000 metric tons of CO_2 are being pumped into a saline aquifer 700 meters below the surface of the Earth. Numerous sensors provide signals that show how the gas is spreading. And the results so far confirm expectations: CO_2 dissolves in the saline water — just as it dissolves in mineral water when a soda water carbonator is used — and is held in the pores of the sandstone layers. After thousands of years, it could lead to the formation of limestone or carbonates. A large amount of the greenhouse gas would then be kept out of the atmosphere permanently.

Subterranean Glass Domes

And what if CO_2 does reach the surface? Since the gas is heavier than air, critics fear that it could collect in pools where it would suffocate all life. In concentrations greater than three percent, the odorless gas causes headaches and dizziness; in concentrations greater than ten percent, it can lead to respiratory arrest. In 1986, the volcanic crater in Cameroon known as Lake Nyos suddenly released 1.6 million tons of CO_2, resulting in the death of 1,700 persons. But if storage sites are chosen properly, there is hardly any need to fear catastrophes of this kind. Above the aquifer in Ketzin, for example, there is a layer of gypsum and clay that seals the arched nine-square-kilometer layer of sandstone like a glass dome. This top layer served the same function previously for a period of more than 40 years, during which the sandstone was used to store natural gas, as is common in many places around the world. In fact, the best CO_2 storage locations would be sites where gases have always been stored underground, which basically means all natural gas reservoirs, because these have demonstrably remained leak-proof for millions of years. Some oil and gas producers already pump CO_2 back into their deposits in order to raise their yield through increased pressure.

Reinhard Hüttl, Scientific Director of the German Research Centre for Geosciences (GFZ) in Potsdam, which is managing the project in Ketzin, points out that CO_2 is generally not toxic. We inhale it in small quantities in the air, or drink it in sparkling mineral water or soft drinks. "But if it really were to leak out, we would notice it immediately with our monitoring systems, and we'd be able to simply disperse it in the air," he says. However, Hüttl concurs that the whole process chain of CO_2 separation, transport, injection, and monitoring is very elaborate and costly. "As a result, CO_2 storage is more of a transitional technology than anything else, as long as we have coal-fired and natural gas power plants," he says.

The Most Important Biochemical Process on Earth

But there is another way to keep CO_2 out of the atmosphere besides storing it underground. It can be combined chemically with other substances. The most intelligent method for doing this was invented by nature itself when, over three billion years ago, the first bacteria learned to use photosynthesis, the most important biochemical process on Earth. Purple bacteria and cyanobacteria were followed later by green, brown, and red algae, and then by green plants. Using energy derived from sunlight, they convert water and the CO_2 in the air into carbohydrates such as glucose and starch. In the process, they release oxygen, which is in turn breathed in by animals and human beings. The coal, oil, natural gas, and biomass that still make up over 90 percent of our energy supply are all based on the results of the plant photosynthesis that took place millions of years ago. And the organisms that carry on photosynthesis are still very active. Every year, they convert roughly seven to ten times the amount of energy consumed by mankind in the same period.

If we look at the details of how plants work, we find that their cells contain special antenna molecules that capture light and convey it to "reaction centers." With their pigments — green chlorophyll and yellow-to-red carotenoids — the antennas cover the spectrum of sunlight very well and can therefore harvest a great deal of light. In the reaction centers, captured photons lead to a separation of positive and negative charges within a few trillionths of a second, thus allowing a plant's "battery" to be recharged in just an instant. The great speed at which this charge separation occurs prevents the positive and negative charges from combining again, which would ultimately result in the loss of the energy of the light in the form of heat.

Instead, plants use their internal battery to decompose water into hydrogen ions and oxygen and build molecules that are rich in energy. Plants then use these energy-rich molecules in combination with airborne CO_2 in order to build the carbohydrates that they need to grow. Ultimately, only the green, photosynthetically active parts of a plant can fix CO_2. At night, however, and in the non-green parts of plants from the roots to the fruit, carbohydrates are broken down again in order to obtain energy in a process that corresponds to our respiration: oxygen is needed for this, and CO_2 is released into the air.

But overall, plants fix more CO_2 than they release. In Germany, a tree incorporates an average of ten to 30 kilograms of atmospheric CO_2 into its biomass per year, depending on its type and size. In tropical regions, it might be two or three times that much. A forest in Central Europe fixes about 1,000 metric tons of CO_2 per square kilometer each year. The extremely fast-growing miscanthus, a type of grass that reaches several meters in height, can fix 5,600 metric tons, which puts it in the lead of all naturally-occurring plants. Only algae have a greater capacity. When grown in bioreactors, they can consume up to 40,000 metric tons of CO2 per square kilometer every year.

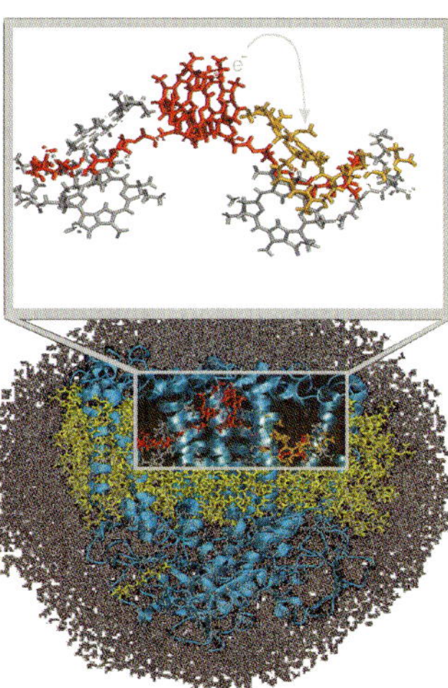

Plants do it, and bacteria do it to. In complex molecules (as in the reaction center of a photosynthetic bacterium, shown at right), they use sunlight to form biomass out of CO2 and water.

Green Glowering Landscapes

It is no surprise that there are already attempts under way to capture the CO_2 from power plant emissions using algae. In 2008, power plant operator RWE Power commissioned a 600-square-meter pilot facility at the Niederaußem lignite-fired power plant outside of Cologne, Germany. A small portion of the flue gas is passed through translucent tubes filled with saltwater in which microalgae swim. They convert roughly 12 tons of CO_2 to biomass each year. However, given the 27 million tons that this giant power plant emits annually, this corresponds to the amount that Niederaußem blows into the air in less than 15 seconds of operation.

Algae love CO_2. They can consume this greenhouse gas directly from power plant emissions, as demonstrated by this test at the Niederaußem lignite-fired power plant.

It is nevertheless a research project that makes sense, of course, considering that algae grow ten times faster than land plants. Furthermore, algae can process CO_2 directly from power plant emissions. Even the waste heat is helpful because it promotes their growth. And the biomass produced by algae can be easily converted into a great variety of useful substances, such as biodiesel, biogas, and bio-plastics, for example.

Biogas and biodiesel could replace the fuels manufactured from petroleum. But the Shell oil company has calculated that biodiesel from algae would be competitive only if the price of crude oil were to exceed 800 dollars per barrel. (One barrel contains about 159 liters.) Eight hundred dollars per barrel is about ten times the current price and still more than five times what crude oil cost at its peak in the summer of 2008.

Nor are biofuels from algae an appropriate solution for keeping CO_2 out of the atmosphere over the longer term, because the release of the original power plant emissions into the air is only delayed, as it were, until the biofuel is burned. Some researchers are therefore pursuing a different idea: Using iron as a fertilizer, they want to stimulate algae growth in the sea in the hope that CO_2 will be removed from the atmosphere when the algae die off and sink to the sea floor. But several experiments in the Indian Ocean and the South Atlantic have shown that this does not work. What researchers hadn't considered was that small crustaceans eat most algae, which means that the CO_2 enters the food chain again and is ultimately exhaled into the atmosphere by animals.

Many scientists also reject these and similar "geoengineering" strategies in principle. When these practices are used, mankind directly intervenes on a large scale in natural biological or chemical processes. In addition to iron fertilization, for example, there are also proposals to use sulfur particles to promote the formation of cooling clouds, or to install solar shields between the sun and the Earth to create vast areas of shade. Regardless of how successful these measures might be, and of whether international agreement could be reached to implement them, the primary points of criticism are the unknown and possibly uncontrollable side effects that such experiments could have.

A different and unquestionably more sensible way of fixing CO_2 would be through planting forests. Forests remove CO_2 from the atmosphere for decades, until the trees that incorporate it into their biomass rot away and release the greenhouse gas again. Forestry experts have calculated that the 11 million hectares of woodland in Germany take in about 17 million metric tons of CO_2 per year. That figure already takes into account the amount of CO2 which is released in Germany as a result of using wood in biomass power plants.

Every year, however, about 13 million hectares of woodland around the world are lost to fires and deforestation, especially in developing countries in tropical regions. Every year, in other words, more woodland is destroyed than Germany itself has within its borders! As a result of this forest loss, six to seven billion tons of CO_2 reach the atmosphere. That's about a fifth of all the CO_2 emissions caused by mankind. The clearing of tropical forests and the burning of peat bogs in Indonesia alone is estimated to produce CO_2 emissions of more than 2.5 billion tons each year – about three times the total CO_2 emissions of Germany.

The associated destruction of biodiversity is also alarming. On a single hectare of Peruvian rainforest, for example, scientists have counted 283 different species of trees, whereas there are only 40 native tree species in all of Bavaria. In total, researchers estimate that tropical rainforests contain approximately 50,000 tree species and about the same number of vertebrate species. Ninety percent of all plant and animal species are found in rainforests. Putting a stop to the over-exploitation of the green lungs of the Earth must therefore be a top priority internationally, considering that the Earth still has about 1.3 billion hectares of tropical forest. At the same time, afforestation must henceforth be counted as a climate-protection measure and be rewarded with carbon credits. The first pilot projects for this are already under way. In 2006, for example, in exchange for reforesting 4,600 hectares of pasture in Costa Rica, Precious Woods, a Swiss company, obtained carbon credits for over 221,700 tons of CO_2 that it was then able to trade on the Chicago Climate Exchange.

A Rock that Transforms CO_2 into Minerals

But the best option for removing CO_2 from the atmosphere may not even come from the world of animate nature. Consider, for a moment, the country of Oman at the eastern corner of the Arabian Peninsula. For several weeks in 2009, it was here that Evelyn Mervine climbed around on one of the largest pieces of rock in the world. Mervine, a geology student at the Woods Hole Oceanographic Institution near Boston, Massachusetts, was studying the Samail ophiolite, a former section of the oceanic crust.

This massive sea of stone is 350 kilometers long, 40 kilometers wide, and five kilometers thick, and it consists mostly of magnesium silicates. "When water penetrates the rock and washes silicates upward, they react with carbon dioxide in the air to form carbonates — in other words, solid minerals, thus removing CO_2 from the atmosphere," says Mervine. "If we can figure out how this process works, maybe it will be possible to accelerate it — that would be an extremely exciting combination of geology and climate protection."

Professor Ron Zevenhoven, who works at Abo Akademi University in Finland, underscores the great potential of this idea. "The rock in Oman alone contains enough magnesium silicates to fix all the carbon from all the fossil fuels on Earth," he says. In nature, this process occurs slowly, but at least it does take place without any sort of external impetus. The formation of carbonates is the only chemical reaction with CO_2 that does not require energy input. Each year, the Samail ophiolite absorbs 100,000 tons of CO_2 dioxide from the air. Researchers hope that they may be able to accelerate this process a hundredfold, by pumping warm water into the rock, for example.

"In principle," says Zevenhoven, "one could simply scatter powder consisting of magnesium silicates — preferably in the warm and wet tropics — and then rain would ensure that these substances weather, or form carbonates with CO_2 in the air." It is difficult to imagine a simpler method of CO2 removal. And magnesium carbonate is a harmless substance that everyone is familiar with. It is used in the form of a loose white powder in climbing sports and gymnastics to give hands a better grip and to dry perspiration. The biggest obstacle to the idea would likely be the sheer quantity of the compound needed. Neutralizing the annual emissions of a large-scale power plant like Niederaußem would require a mountain of magnesium carbonate measuring about one square kilometer at its base and extending over 17 meters high — not to mention the expense necessary to obtain the huge amounts of silicates from Oman and transport them.

But there are many other ways of removing CO_2 through its uses in the chemical industry. It is used to produce urea — an important nitrogen fertilizer — or for salicylic acid, from which aspirin is derived. It is also used for firefighting, as a refrigerant, and as carbonic acid in beverages. About 80 million tons of CO_2 is used each year for urea production alone. In all, the DECHEMA Society for Chemical Engineering and Biotechnology estimates that the global chemicals industry could use a maximum of two billion tons of CO_2 per year. It would probably not be economically and technically feasible to use more than about 200 million tons, however.

Converting CO_2 into Fuel

It would make very good sense to replace oil as a fuel, because there is hardly any less sensible use of petroleum than burning it in the engines of cars, where only 20 to 30 percent of its energy is actually converted to locomotion, while the rest is lost as heat. Possible substitutes for oil include not only the biofuels mentioned previously but also the alcohol methanol. For years, the American Nobel laureate in chemistry George A. Olah has been propagating the idea of a methanol economy to replace the petroleum era. Methanol can be stored with relative ease in tanks currently used for gasoline. They would have to be somewhat larger, and some parts with which methanol reacts aggressively would have to be redesigned, but otherwise the same infrastructure could be used, from filling stations to cars.

And the nicest part is that like methane, which is contained in natural gas, methanol can be produced from hydrogen and airborne CO_2. When methanol is burned, it produces no additional CO_2 emissions. The CO_2 that was previously part of the air is merely released back into the atmosphere. This assumes that the hydrogen needed is produced regeneratively, in other words by electricity generated from wind power or solar cells, for example. But it remains to be seen whether all of this makes sense economically. Large facilities for decomposing water (i.e. electrolyzers) have efficiency ratings of between 60 and 70 percent. The resulting hydrogen therefore has 30 to 40 percent less energy than the electricity put into the process. About one fifth of the energy is also lost during methanol synthesis.

When everything is considered, this means that the sum total of current wind energy in Germany could only produce enough methanol to supply seven percent of the country's passenger cars. On the other hand, however, the same quantity of wind energy would suffice to power four times as many cars for an entire year, provided they were electric cars.

Learning from Plants

No matter how you look at it, scientists who are searching for ways to keep CO_2 out of the atmosphere will have their work cut out for them in coming decades. The best thing would be for them to work all the angles at once: Remove CO_2 from power plant emissions, build "green" power plants, optimize the conversion of algae into biofuels, put efficient electric cars on the road, find effective methods for storing electricity — and above all, learn from nature. After all, both with regard to its importance for life on Earth, and purely from a technical point of view, there is probably no process more ingeniously designed than photosynthesis.

Almost every photon captured by biological antennas is used; hardly anything is lost. The first step, which leads to the separation of positive and negative charges and thus recharges a plant's inner battery, is among the fastest chemical processes ever measured. It takes place in large molecules, the reaction centers, in which more than ten thousand individual atoms are so perfectly attuned to one another that they waste a minimum of energy. These small light converters measure only a few millionths of a millimeter in diameter. They are thus tinier than the smallest components that researchers can manufacture with current nanotechnology – and at the same time, their structure is much more complex. Nevertheless, these complex structures continually recreate themselves on their own in a clockwork of biochemical processes – without any engineer to intervene. It is a perfectly self-organized process.

For decades, the dream of many photosynthesis researchers has been to replicate these systems with human engineering and adapt them to human needs. Plants have not optimized their cellular apparatus for the energy needs of human beings but rather for their own growth and reproduction. Although a sort of "nanobattery" is created in a plant cell and hydrogen ions are formed, for instance, it makes little sense to try to turn this into a solar battery, or to interrupt the process in such a way that plant cells or algae produce hydrogen in large quantities.

Instead, scientists must try to understand in detail how these biomolecules operate, so that they can then apply this knowledge to the optimization of technical systems, whether it be in solar cells, hydrogen production plants, or molecular high-speed switches. And researchers have made some progress along this path since the 1980s and 1990s, when the principal processes of photosynthesis were revealed and the findings were rewarded with several Nobel Prizes. They have succeeded in manufacturing artificial relay races of electrons that move the electrons along very rapidly and thereby accomplish a separation of charges almost as good as that found in the photosynthetic reaction center.

In another instance, biological antennas have been replicated by research groups. Chemists at the University of Würzburg, Germany, have assembled thousands of molecules to form nanocapsules. The molecules within the capsules act like antennas: They capture light and transfer the energy to other molecules.

Colored Solar Sells in Backpacks

In the early 1990s, Michael Grätzel in the Swiss city of Lausanne invented a solar cell that also incorporates principles of photosynthesis. It uses a porous layer of titanium dioxide nanocrystals only a hundredth of a millimeter thick with an adjacent layer of electrolyte solution sandwiched next to it. Dye molecules are bonded to the surface of the titanium dioxide particles. When light reaches a dye molecule, the effect is quite similar to what happens in photosynthesis. The dye captures the light and immediately passes on the energy, in this case, to the tiny particle of titanium dioxide in the vicinity. That particle in turn releases an electron that travels across the network of other titanium dioxide particles until it reaches an electrode. A flow of current is thus generated.

These "dye-sensitized" solar cells are easier and cheaper to produce than traditional cells made of silicon. They currently attain an efficiency of 12 percent, which is half that of conventional silicon solar cells. But they have a further advantage. They can be packaged in plastic, which makes them flexible. Since late 2009, in fact, G24 Innovations, a company headquartered in Cardiff, Wales, has been using a simple process to manufacture rolls of shimmering, reddish-tinged "Grätzel cells." A company in Hong Kong is already integrating them into backpacks and plans to use them in tents too, where they can supply energy to cell phones, e-books, cameras, and LED lights.

Solar cells made of plastic. Unlike silicon solar cells, dye-sensitized solar cells can even be produced in rolls of flexible plastic film.

Other researchers hope that by studying photosynthesis, they will learn how water can be decomposed as efficiently as possible using only sunlight. They want to link antennas for capturing light with molecules that cause a flow of electric current. By means of attached catalysts, that current could in turn be used to produce hydrogen and oxygen from water — or to create methanol directly with CO_2 from air.

Reaching this objective will take time, and there are many obstacles to overcome. For example, the decomposition of water gives rise to aggressive substances. Plants solve this problem by continually repairing and renewing their green catalysts. A technological imitation cannot do this; it must rely on catalysts that are more stable. Scientists, such as those at the Jülich Research Center in Germany, have developed some initial ideas for solving this problem, but a great deal of painstaking, detailed work will have to be done before mankind is capable of using the rays of the sun as ingeniously as plants can.

Monte Carlo, August 2050. Ancient Rome is this month's theme at the new casino. Staff members are wearing the magnificent armor of centurions, complete with brush-maned helmets, and at the reception desk togas and tunics are available for the guests. But when it comes to energy consumption, the casino is state-of-the-art. Thanks to solar collectors on the roof and smart efficiency-enhancing technologies, it has reduced the amount of electricity it draws from the power grid by 80 percent. A nanocoating on the windows allows solar radiation to heat the interior only as needed. Power-saving organic LEDs provide glare-free light, and microsensors distributed throughout the complex ensure optimal air conditioning. Only the most efficient types of refrigerators, stoves, and washing machines are used. Computers and 3-D displays don't need any power at all in standby mode, and even the suburban railroad has reduced its already low power consumption by a third, thanks to lightweight construction materials and braking energy recovery. Next year the casino management plans to buy its own small wave power plant, which will convert the forces from coastal tides into electricity. At that point, the casino will finally generate more energy than it consumes. The hotel management is even planning to earn money by selling the surplus electricity on the utilities exchange.

THE POWER OF THRIFT: FROM MEGAWATTS TO NEGAWATTS

Capturing carbon dioxide and utilizing it by technical means is helpful, but it will not be sufficient in the struggle against climate change. There will be no simple solutions — every technology will have to contribute to the effort. If we compare the scenarios of the International Energy Agency, Shell, the German Federal Ministry for the Environment, and Greenpeace, the emphases are somewhat different depending on their respective political orientations, but essentially all the trends point in the same direction. In order to cut worldwide CO_2 emissions in half by 2050, experts estimate that:

- about a third of the CO_2 reductions can be achieved through the intense expansion of renewable energies. Experts at Greenpeace even believe that by 2050 more than half of the world's energy needs could be covered by renewable energy sources.
- about one fifth of the necessary CO_2 reduction can be achieved through the separation, storage or use of carbon dioxide,
- approximately half of the CO_2 reduction must come from higher energy efficiency in the production, transmission, and use of energy in industry, households, and transportation.

The tremendous efficiency potential of power plants has already been described. If state-of-the-art technologies were comprehensively implemented, power plants fired by coal and natural gas could save billions of tons of CO_2 per year. Germany has already achieved impressive levels of energy efficiency, as can be seen from a brief calculation. The ratio between a country's gross domestic product — adjusted to take purchasing power into account — and the energy it consumes is a rough measure of how efficiently that nation utilizes energy. If we set the value for Germany at 100, the U.S. and China have values of about 70 and Russia has 33. In other words, if the U.S., China, and Russia managed to use energy as efficiently as Germany, that alone would reduce worldwide energy use by 15 percent.

The situation is similar for carbon dioxide emissions. Here the values are 147 for the U.S., 179 for China, and 291 for Russia. In other words, for the same amount of GDP, the U.S. generates 47 percent more CO_2 than Germany, China 79 percent more, and Russia 191 percent more. If these three countries succeeded in reducing their relative emissions to Germany's level, global CO_2 emissions would decline by approximately 20 percent.

And these are only the values for the use of technologies that Germany employs today. But the future potential is even greater, for even in Germany much more can still be done. U.S. physicist Amory Lovins, a managing director of the Rocky Mountain Institute for sustainable development, believes that energy conservation is the biggest energy resource of all. Back in the 1970s Lovins, the recipient of the Right Livelihood Award, invented the concept of "negawatts" — the megawatts saved through higher levels of energy efficiency. Here, the idea is that it is often cheaper and simpler not to build real power plants, but to use electricity more efficiently. The resulting energy savings can add up to the equivalent of what a real power plant would consume to power thousands of inefficient homes and appliances.

Doubling Wealth while Cutting Resource Use in Half

In the 1995 book Factor Four, which he wrote together with his then-wife and the physicist Ernst Ulrich von Weizsäcker, Lovins called for methods of doubling prosperity while halving the use of natural resources. He pointed out that many environmentally-friendly innovations can increase energy efficiency at least fourfold. Lovins is currently working on a program he calls "Reinventing Fire." "We're developing a comprehensive roadmap for a profitable transition from coal and oil to efficiency and renewable energy," he says.

Lovins describes our current energy supply as a pyramid whose base consists of coal-fired and nuclear power plants. In the middle is natural gas, and at the top are renewable energy sources. He calls for nothing less than turning this pyramid upside down in the coming decades, so that "the base is far more efficient electricity consumption. Renewable energy sources and some combined heat and power make up the middle. At the top of the pyramid, the remaining fossil fuels and nuclear power will gradually be phased out, in much the same way that we phased out steam locomotives."

Many examples show how right Lovins is to focus on energy efficiency. For instance, energy-saving bulbs and LEDs produce the same degree of luminous efficiency using only one fifth of the electricity needed by conventional light bulbs. That's why legislators in many countries have decided to gradually replace incandescent bulbs, especially since the new light sources' life spans are also 15 to 50 times longer. Almost 19 percent of the electricity consumed worldwide is still used for lighting alone; given today's energy mix, that generates two billion tons of CO_2 emissions annually. Thanks to energy-saving lighting, this amount could be drastically reduced. To use Lovins' terminology, between 200 and 300 medium-sized coal-fired power plants could be closed down throughout the world and replaced with "negawatt power plants."

Researchers at Osram have created an ecological balance sheet after taking a close look at the life cycle of light bulbs — from the quarry where the basic raw materials for glass are mined all the way to the recycling facility or landfill. This has revealed that only one to two percent of the total energy use associated with a light bulb occurs during production, and more than 98 percent occurs during operation. This means that even a single energy-saving bulb delivers impressive performance. Because of its reduced power consumption, it saves approximately half a ton of CO_2 throughout its entire life cycle — more CO_2 than a tree can bind in the same period of time. And it also saves money. At today's electricity prices, it recoups its additional cost after only a few hundred hours of operation. Throughout its entire life cycle of about 15,000 hours, an energy-saving bulb saves more than € 200.

In view of these figures, it's obvious why many cities are replacing the incandescent bulbs in their traffic lights with even longer-lasting LEDs, which also use 80 percent less power and have to be replaced only once a decade. Such an investment is certainly worthwhile. It enables a city with 700 intersections to save €1.2 million a year in energy costs. For Germany as a whole, the reduction in power consumption alone would reduce the cost of its traffic lights by €140 million a year.

For home appliances as well, the operation phase accounts for 90 to 95 percent of the overall environmental impact. Here too, low energy use is vital for a good ecological balance sheet. Replacing old appliances is therefore worthwhile, because refrigerators, freezers, washing machines, dryers, and other kitchen appliances taken together account for almost half the energy use of German households — and new appliances are significantly more energy-efficient. Compared to the 1990s, their energy consumption has been reduced by 30 to 80 percent!

A similar "power diet" has been prescribed for high-tech equipment such as TVs, stereo systems, and computers, which currently account for one fifth of German households' energy consumption. For example, energy-efficient TVs — such as LCD TVs, which use LEDs for background lighting — require only half as much power as large conventional TVs. It's also very important to reduce or completely eliminate the standby operation of electric devices. There are already computer monitors that require no power at all in standby operation, because a built-in capacitor stores enough energy over several days to be able to switch on the display again when the computer gives the corresponding signal.

Of course there are also tremendous potential energy savings in industry as well. For example, about 20 million large electric motors are being used throughout the world for pumps and drive systems of every kind, such as those used for conveyor belts. Such motors account for around two thirds of the total power consumption in industry. In Germany, industries consume one and a half times as much power as all of the country's households combined. But a great deal can be done to change that. The power consumption of many pumps and drive systems can be reduced by up to 60 percent through the use of more efficient and intelligently-controlled motors that don't rotate at a constant speed, as they do today, but instead adjust their rpm to the task at hand.

A similar amount of potential for modernization is offered by iron and steelworks, which devour up to a fifth of the energy required by industry and generate 30 percent of industrial CO_2 emissions. That's why here too energy-efficient technologies can combat the economic crisis and the climate crisis simultaneously. One example is steam turbines, which use the steam arising from the cooling of coke to generate electricity; the use of steam turbines in a single coking plant can produce enough additional power to run approximately 30,000 households. Waste heat can be used in many places, for example to generate power in the glass, metal, and cement industries. Airports also offer tremendous potential. For example, efficient energy management and the optimization of heating, cooling, ventilation, lighting, and baggage handling systems can result in energy savings of as much as 40 percent.

Financing Investments through Energy Savings

Cost savings of a similar magnitude can be achieved through smart energy reduction measures in hospitals, universities, schools, swimming pools, and many other types of buildings. Two good examples are the clinic in Bremerhaven, which has reduced its energy costs by more than €500,000 annually, and 11 municipal swimming pools in Berlin, which have saved over €1.6 million. Here the operators replaced old boilers, converted from oil to gas operation, and installed more efficient heat recovery and warm water processing systems. If all of the world's office buildings, hospitals, schools, and universities were renovated in ways that resulted in energy savings of about 30 percent, total CO_2 emissions would decrease by about 500 million tons a year — the total volume of emissions the UK generates today.

Particularly interesting for municipalities is an innovative method of financing such measures called energy performance contracting. This means that municipalities don't need to pay any money up front — the savings resulting from the modernization measures are contractually guaranteed, and the municipalities can repay their investments in installments with part of the energy costs they save over a period of ten years, for example. If the savings are greater than the installments to be paid, the surplus remains in the municipal treasury and can be used for further projects. This model is so attractive that it is now being used all over the world, not only for thousands of public buildings but also for the conversion of traffic light systems to LEDs and other measures. What all of these measures have in common is the fact that they lead to energy cost savings that are so high that they alone can pay for the original investments in just a few years. This is exactly what Lovins means when he says it's more worthwhile to invest in negawatts than in megawatts.

The financing dilemma faced by municipalities is basically the same one that prevents many people from buying energy-saving lamps or LEDs. These initially cost significantly more than cheap energy guzzlers such as incandescent light bulbs. The increase in the initial investment obscures the fact that it saves the investor lots of money in the long term. This reluctance is even greater when it comes to the modernization of private homes, whether it's a question of wall insulation, improved window insulation or the installation of a new boiler. The homeowner must first make an investment before he or she can profit from it over the years — and there may not be a profit at all, as in the case of apartment buildings. That's because lower energy costs reduce the tenants' utility fees. The owner doesn't benefit unless he or she raises the rents or can at least pass part of the renovation costs on to the tenants.

Such financial obstacles must be overcome more effectively in the future so that private homeowners can also play a role in the green energy revolution. To this end, proactive measures by governments and municipalities are needed, as are innovative banks that offer concepts such as energy performance contracting for private customers as well. If these measures work, then one of the biggest areas of potential energy savings could be

Efficient lighting is an essential aspect of modern building technology. Here, 12,000 LEDs illuminate the arch over the new stadium in Durban, South Africa.

tapped by 2050, because buildings account for approximately 40 percent of worldwide energy consumption and more than a fifth of all greenhouse gas emissions — about 9.5 billion tons of CO_2 in all. Almost six billion tons of this are indirect emissions resulting from electricity consumption in buildings, while 3.5 billion tons are direct emissions due primarily to heating systems using oil or natural gas. Both types of emission can be significantly reduced through more efficient appliances as well as heating systems that in some cases generate no emissions at all — as will be shown in the next chapter.

Why Guidance Systems Boost Transport Efficiency

In addition to buildings and industry, transport — the third major area of energy consumption, which generates 5.5 billion tons of CO_2 worldwide annually — also offers huge opportunities for enhancing efficiency. For example, thanks to lightweight construction materials and braking energy recovery, new buses and trains consume 30 to 40 percent less energy than older models. Vehicles with hybrid technology also consume up to 30 percent less fuel. Equally thrifty are cars with the latest diesel engines that require only 3.8 liters of fuel per 100 kilometers, which corresponds to CO_2 emissions of 99 grams per kilometer. Even more efficient vehicles consuming two to three liters of fuel per 100 kilometers — for example, diesel-electric hybrid vehicles — are expected to enter mass production within the current decade. On the basis of today's mix of power sources, purely electric cars would generate about 90 grams of CO_2 per kilometer. If electricity from renewable energy sources such as wind and the sun were used, their CO_2 emissions would be reduced to almost zero. Traffic guidance systems can also help to boost transport efficiency, because they reduce congestion and make it easier to find parking spaces. Studies have shown that up to 40 percent of urban traffic consists of drivers looking for a place to park.

Railroads are also very environmentally friendly. High-speed trains like those in Spain and China, which reach speeds of 300 to 450 kilometers per hour, offer power consumption figures that correspond to 0.33 liters of gasoline per passenger per 100 kilometers. This is an overly positive calculation because it does not take into account the emissions generated by today's power plants — but even if the emissions are added in, the power consumption per passenger is just over one liter per hundred kilometers. For long-distance trains in Germany with a typical passenger load, the German railroad Deutsche Bahn calculates that this figure is currently just over two liters per passenger. And even that is quite a bit better than airplanes, which need about three to four liters of kerosene per passenger per 100 kilometers.

Airplanes also have an additional problem because they emit not only CO_2 but also nitrogen oxides, which promote the formation of ozone. Worse yet, they generate steam, which eventually forms vapor trails and cirrus clouds — clouds of ice floating high up in the atmosphere — both of which reflect back the heat coming from the earth, thus intensifying the greenhouse effect. As a result, the impact of air traffic on the climate is at least three times as great as a comparable emission of these pollutants on the ground, according to the German Federal Environment Agency (UBA). Experts at the UBA estimate that because of these emissions of CO_2, nitrogen oxides, and steam, airplanes today contribute about as much as worldwide automotive traffic to the greenhouse effect. The International Air Transport Association (IATA) has responded to this. In 2009 it decided to implement numerous measures such as new types of jet engines and fuels, the optimization of flight routes, and air traffic monitoring in order to reduce the CO_2 emissions of air traffic to half of the 2005 level by 2050.

Everyone Counts

Whether we're looking at traffic, industry, or buildings, the overall opportunities for using energy more effectively are tremendous, and they promise gigantic markets for all the companies that operate in these sectors. This greatly increases the likelihood that we will actually be enjoying the benefits of "negawatts" by 2050. This approach will pay off in any case, because even the people who doubt that climate change is taking place do not deny that our fossil resources are growing scarce. For this reason alone, energy prices will increase drastically in the future — so if we manage to get more energy from the raw materials that remain we will be saving money.

Above and beyond that, we need to focus primarily on transforming the principle of sustainability into a categorical imperative in people's minds, as the philosopher Immanuel Kant might have put it. "Always act in such a way that future generations will also have a world that is worth living in." This 21st-century principle is the essence of sustainability. We must succeed in making the preservation of the basis of life on Earth a principle of universal ethics — just as we have done for human rights, which are recognized by almost every nation, even though these rights are interpreted differently by some of them. The protection of nature and the careful use of the Earth's resources — which include raw materials, drinking water, and the atmosphere's capacity to absorb greenhouse gases — must become a matter of course in all of our actions. That's because ultimately every individual's contribution counts, whether it's a matter of buying power-saving appliances, doing more recycling, turning solar cells on the roof and electric cars in the garage into status symbols, condemning the wasting of energy, making the use of bicycles and public transportation a habit, compensating for air travel through reforestation certificates — or simply eating less meat, which would reduce methane gas emissions from farm animals, which according to a UN study account for more than 18 percent of worldwide greenhouse gas emissions.

Ernst Ulrich von Weizsäcker has emphasized the opportunities that technology offers to transform the energy system, an event that would be unique in the history of mankind: "Today we can conjure up ten times more light from a kilowatt-hour of electricity than just a few years ago. Buildings can be kept warm with a tenth of the heating energy used back then. The whole country could become five times as energy-efficient as it is today." But he adds, "However, as long as energy remains cheap, that will not happen." Only if people are offered the right incentives will they change their behavior.

According to Weizsäcker, we should therefore not hesitate to make energy use more expensive via taxes or CO_2 emission certificates — "in small steps parallel to growing energy efficiency, so that investors can make long-term plans." This would also be fair in social terms, and poorer households could also receive state support. Weizsäcker is optimistic that the energy revolution can succeed: "Our cars, houses, and appliances are ungainly wasters of energy that have become outdated. But we will manage to get back on track. I see a society coming that is more efficient and more elegant than the one we have today. That won't entail a drop in the quality of life — on the contrary, I see us setting out for a new age of advanced civilization." That's a vision of a civilization that in 2050 will no longer be based on the exploitation of nature and the combustion of coal and oil but on significantly smarter and more diverse solutions that deal more sparingly with natural resources and the environment.

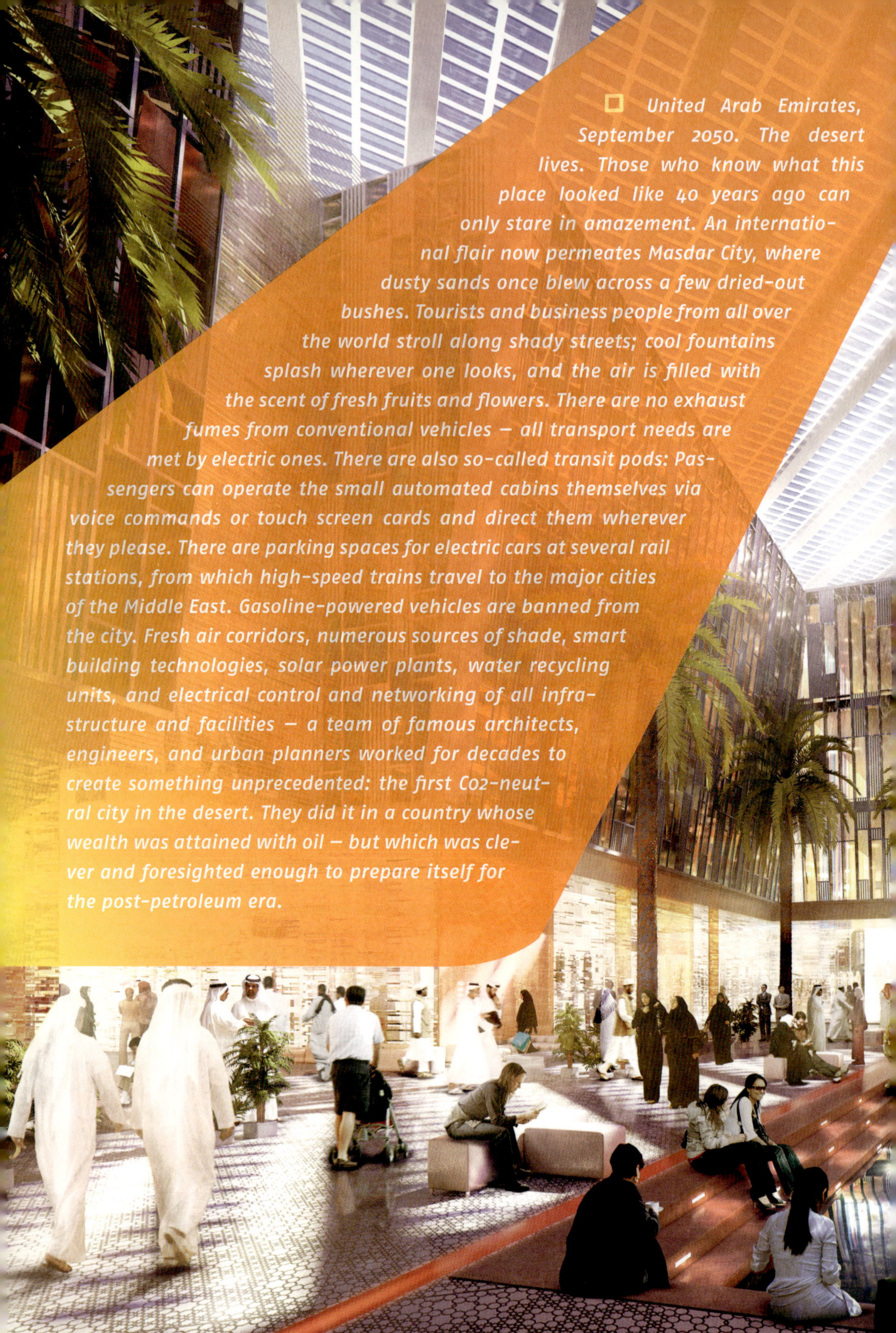

United Arab Emirates, September 2050. The desert lives. Those who know what this place looked like 40 years ago can only stare in amazement. An international flair now permeates Masdar City, where dusty sands once blew across a few dried-out bushes. Tourists and business people from all over the world stroll along shady streets; cool fountains splash wherever one looks, and the air is filled with the scent of fresh fruits and flowers. There are no exhaust fumes from conventional vehicles — all transport needs are met by electric ones. There are also so-called transit pods: Passengers can operate the small automated cabins themselves via voice commands or touch screen cards and direct them wherever they please. There are parking spaces for electric cars at several rail stations, from which high-speed trains travel to the major cities of the Middle East. Gasoline-powered vehicles are banned from the city. Fresh air corridors, numerous sources of shade, smart building technologies, solar power plants, water recycling units, and electrical control and networking of all infrastructure and facilities — a team of famous architects, engineers, and urban planners worked for decades to create something unprecedented: the first CO2-neutral city in the desert. They did it in a country whose wealth was attained with oil — but which was clever and foresighted enough to prepare itself for the post-petroleum era.

THE ZERO-EMISSION CITY

Munich in the year 2050 is as green as can be — and not only because of its many parks and its close proximity to the alpine upland. The German metropolis also generates its energy in an environmentally-friendly way, and because its residents use this energy so efficiently, each of them produces only 750 kilograms of carbon dioxide (CO_2) per year on average — or 90 percent less than today. This figure is also well below the two-ton maximum stipulated for each human being in 2050 as part of the effort to limit the global temperature increase to two degrees Celsius. Munich is therefore practically emission-free, an outstanding achievement for a non-planned city. Buildings hardly need to be heated here any more, and those that do require heating get it through mini-power plants operated by their owners, or else from district heating systems. Electricity is produced at solar, wind, or geothermal plants, and the streets are filled with quiet electric vehicles. The above description is not a utopian vision — it's a realistic scenario, according to researchers at the Wuppertal Institute for Climate, Environment, and Energy.

The institute's study Munich — A Roadmap to a CO_2-free Future, which was published in 2009, looks ahead to the year 2058, analyzing in detail which measures will achieve the greatest reduction of CO_2 emissions and whether they are economical. For example, heating requirements for Munich's buildings alone currently accounts for nearly 46 percent of the Bavarian capital's CO_2 emissions. Older buildings should therefore be renovated in accordance with the so-called passive house standard, and new construction should conform to it at the very least. This includes the use of not only the very best insulation for walls and ceilings, but also vacuum-insulated windows and ventilation systems that keep heat inside a house before blowing air outside. The heating requirement in an older building renovated in this way decreases from approximately 200 kilowatt-hours per square meter per year (kWh/m²a) to 25–35 kWh/m²a. A new building that complies with the standard will consume only 10 to 20 kWh/m²a, or well below the value stipulated in Germany's most recent energy-saving ordinance from October 2009, which sets a maximum new-building heating requirement of about 50 kWh/m²a.

Spend 13 Billion Euros, Save 30 Billion

To ensure that virtually all buildings in Munich can be equipped for this during the next 50 years, the rate at which refurbishments are carried out must increase from the current figure of 0.5 to 2 percent per year. In other words, every year four times as many homeowners will have to undertake such improvements than is currently the case. That's a very ambitious goal, but it's also feasible and it literally pays off. It's true that the passive house standard generates costs higher than those associated with a building that "only" conforms to the energy-saving ordinance. According to the Wuppertal Institute, renovations and new passive house constructions would generate around € 13 billion in additional costs as compared to compliance with the 2007 ordinance. That amounts to € 200 a year per resident, or around one third of their current annual gas bill. However, such an approach would ultimately generate between € 1,200 and € 2,000 in energy savings per person per year. Conclusion: Consistent adherence to the passive house standard in renovation and new construction would result in total energy savings of more than € 30 billion between now and 2058.

The Wuppertal Institute also recommends many other measures — for example, the use of extremely efficient cogeneration power plants that produce both electricity and heat. Geothermal plants, such as those already being built just outside of Munich, could also generate both electricity and heat, with the latter sent on to households via district heating pipes. The adoption of the institute's approach would mark the end of individual household gas and oil heating in Munich, which currently covers as much as 75 percent of the city's building heating requirement. Buildings in the future would then require only one fifth as much heating as they need today; the rest could easily be supplied by cogeneration plants, which would result in an increase in the share of heat accounted for by district heating from 20 percent today to 60 percent. This scenario isn't unrealistic. Copenhagen's district heating system already covers 70 percent of all households, for example; in Mannheim the figure is 60 percent, and in Flensburg it has even reached 90 percent.

The institute's scientists also predict that large centralized power plants will produce less and less electricity in the future. Instead, power will be generated more frequently in distributed systems — for example, in cogeneration plants located directly in the neighborhoods they serve, or even in thermal gas units that generate both heat and power directly in households. If the Munich of the future consistently exploits all the electricity conservation potential in everything from traffic lights to dryers, the lion's share of the city's electricity requirement could be generated from CO_2-free sources by 2050. Such sources will include photovoltaic units on roofs and building façades, small wind power plants, and geothermal facilities. Nevertheless,

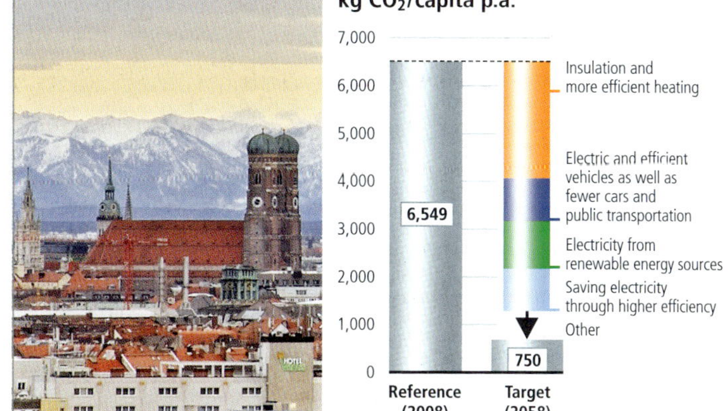

Ninety percent less CO₂. Today's technology could dramatically reduce Munich's CO₂ emissions by 2058 without any sacrifice in quality of life.

a portion of the energy will still have to be produced at large power plants — but these could also take the form of extensive wind parks in northern Europe or major solar power plants in southern Europe or North Africa from which the green electricity would be transported to Germany with very low losses.

Munich's power utility (SWM) operates ten hydroelectric plants. The company also has an interest in large wind parks on the North Sea and in the state of Brandenburg as well as solar power plants in Germany and Andalusia. In December 2009 the city of Munich announced that SWM planned to invest around € 9 billion between then and 2025 in order to achieve a goal that would make Munich unique among major German cities: the generation of more than two terawatt-hours (TWh) of environmentally friendly electricity per year at its own facilities by 2015, enough to supply all 800,000 private households in the city. That figure would then increase to 7.5 TWh of green power by 2025 — the equivalent of Munich's total electricity requirement in December 2009.

With regard to transportation, the institute's study recommends further expansion of local public transport networks, which should also be made more comfortable in order to encourage more people to use trains and buses. Possible measures here include individual information services for mobile devices that would give passengers updates about public transport connections and thus ease the transition. The study also examines the use of electric vehicles (see p. 125 ff.). For example, it is highly conceivable (and would also make good sense) that by the middle of the century most car trips within Munich will be made in electric vehicles. For longer trips, people will use hybrid cars or efficient diesel and gasoline-powered vehicles. The researchers would also like to see a reduction of traffic through more targeted urban planning — "a city of short distances," as they call it. In other words, instead of building shopping malls at green field sites, new urban districts should be created in which residences, workplaces, and stores are all located near one another. That way, many more trips could be made on foot or by bicycle.

Europe's Greenest Cities

Many of the results of the Munich study can be applied to other cities as well. According to a study carried out for London by the McKinsey consulting firm, for example, technology already available today could be used to reduce energy and water consumption as well as waste and CO_2 emissions by 44 percent by 2025 without negatively impacting the lifestyle of the city's residents. The necessary investment would be less than one percent of London's annual economic output. Most of the 200 sub-solutions identified by the study also make economic sense because they help save energy and money.

A study by the Economist Intelligence Unit that was presented at the global climate summit in Copenhagen in December 2009 examined environmental activities and results in 30 major European cities. The European Green City Index produced by the study shows that in most cases the cities with the best quality of life are also the most advanced in terms of environmental protection. Copenhagen, Stockholm, and Oslo currently head the index, followed by Vienna and Amsterdam. Berlin is ranked eighth — but it's first in building modernization. That's because since 1990 the German capital has succeeded in reducing the average energy consumption of all its buildings by nearly 50 percent, from 150 kWh/m^{2}a to 80 kWh/m^{2}a.

Environmental protection has been a major priority in Scandinavia for many years, and Copenhagen now plans to become completely CO_2-free by 2025. The Scandinavian countries' prosperity is above average and they use the financial leeway to invest in environmental protection. However, it's also interesting to note that nearly all the cities in the study had already developed environmental strategies, and that even some Eastern European cities such as Warsaw, Bratislava, and Kiev were able to successfully implement several strategy elements. In this sense, the study illustrates the heightened awareness of these issues in city halls throughout Europe.

Many cities have launched amazing campaigns. Prague, for example, is using income generated from CO_2 emission certificates to provide financial support to residents who want to modernize their houses. Istanbul is equipping its landfills with four facilities that obtain methane gas from waste and then convert the gas into electricity for 100,000 households. Buses in Amsterdam are now running on fuel produced from garbage, Brussels is funding private carpools, and Dublin has made bicycle purchases tax exempt for people who use them to commute. Tallinn is equipping its buses with devices that cause traffic lights ahead to turn green more quickly, and Ljubljana has set up a lottery for recycling bins in which people who separate their garbage correctly have a chance to win cash prizes.

Such measures explain why nearly all of these cities have per capita CO_2 emissions below the EU average of 8.5 tons per year. Many of them also benefit from the fact that they have very few industries that consume a lot of energy and generate high levels of pollution. "Nevertheless, all of the cities face major challenges," says James Watson, who drew up the European Green City Index. "At the moment, renewable energy only accounts for about 7 percent of their total energy production, for example." Less than a fifth of their total waste is recycled and much of their infrastructure — whether for energy, water supply, telecommunication, or transport — is in urgent need of modernization. A study conducted by Morgan Stanley Investment Management estimates that Europe has to spend € 4.8 trillion on water supply modernization between now and 2030, an additional € 1 trillion on energy supply modernization, and some € 3 trillion on road and rail improvements.

Two Billion New City Dwellers

Europeans may think they are facing tough challenges, but they are living in a land of bliss. Asia will have to spend twice as much as Europe on water infrastructure modernization between now and 2030, and four times as much (more than € 4 trillion) on energy infrastructure — a huge amount of money that most of the region's nations, with the possible exception of China, will have a major problem coming up with. Still, the focus on cities is the right approach, according to Frauke Kraas, a professor and urban studies expert at the University of Cologne. "Humanity's opportunities, problems, and perspectives will all be decided in the cities," she predicts.

This should have been clear to everyone since at least 2007, "because that was the year when Homo sapiens became Homo urbanus," says Anna Kajumulo Tibaijuka, a professor of economics from Tanzania who is the director of the United Nations Human Settlements Program (UN–HABITAT). The year 2007 marked the first time in human history that more people lived in cities than in rural regions. Urbanization has become a megatrend — an irreversible development that affects the entire planet. Whereas not even 30 percent of the world's population could be considered urban in the 1950s, this proportion will probably double by 2030 and reach nearly 70 percent by 2050.

That level has already been exceeded in the industrialized nations, where approximately three quarters of people now live in cities — a figure that is stabilizing at this high level. The same can not be said about the developing countries, however. "Their cities will have to absorb the equivalent of virtually the entire growth in world population between 2000 and 2030, or around two billion people," says Kraas. The major cities of Asia and Africa alone will need to accommodate some 1.5 billion new residents during this period. Experts predict that in 2015 there will be around 560 cities with more than a million inhabitants and that by then 350 million people will be living in megacities with more than ten million residents. Just 30 years ago there were only four cities like that: New York, Tokyo, Shanghai, and Mexico City. Today there are 21.

Living in Berlin, Working in Hamburg

Large cities are also growing closer together. High-speed trains in China, Japan, and Spain that can travel between 300 and 450 km/h are already giving the airline industry a run for its money, and trains are likely to become the better mode of transport for distances up to 1,000 kilometers in the future. China alone plans to build more than 40 high-speed rail routes over the next few years to link most of its megacities. It will then have more of these lightning-fast trains than the rest of the world combined.

According to urban studies experts, any place that can be reached easily within an hour should be considered part of a city's catchment area. In the Middle Ages, when travel was limited to walking or horse-drawn carriages, that would amount to only a few kilometers. Today cars and trains increase that range to between 40 and 50 kilometers — and in the future it could extend to over 200 kilometers. High-speed trains would then make it possible to live in Berlin and work in Hamburg, for example, or take a quick shopping trip from Cologne to Brussels.

Residents of huge cities like Tokyo, Buenos Aires, and Bangkok already produce between 35 and 45 percent of their countries' entire economic output, which explains why urban centers are so attractive. People who move there expect to find jobs and better opportunities to attain affluence than in the countryside. Major cities also offer better medical care, more education opportunities, greater cultural diversity, a more comfortable living environment, and more efficient access to knowledge and communication via the Internet. On the other side of the coin are the huge problems that rapid urban growth creates. In some cases, traffic can come to a virtual standstill — for example in Beijing, where one thousand new passenger cars are registered every day, with dramatic consequences. Over the past ten years the average speed of traffic, even on the city's major ring highways, has declined from 45 to just 10 km/h.

The situation is similar in Bangkok, Mexico City, and Moscow. The Russian capital now has nearly four million registered vehicles that move along only a little faster than pedestrians during rush hours. Some businesspeople in large cities like São Paulo even stay off the road for days at a time, preferring instead to fly by helicopter from their apartments to the roofs of their office buildings. This is partly due to security considerations, of course, but it's also because many streets have become impassable. It's therefore not surprising that a survey of 600 politicians, urban planners, business leaders, and scientists from 25 cities with more than a million inhabitants found that almost half of them believe that traffic poses the biggest challenge in their cities — one far greater than inefficient infrastructures, air and water pollution, and garbage and wastewater systems.

Smog over Beijing. Here, some 1,000 new cars are registered every day, and traffic regularly comes to a standstill as a result.

Cities already consume 60 percent of all drinking water worldwide, either directly or indirectly through irrigation for food crops. Moreover, although cities occupy only 1 percent of the Earth's surface, they account for an estimated 75 percent of global energy consumption and 80 percent of greenhouse gas emissions generated by human activity. Still, urban centers will be hard hit by climate change. Shanghai, for example, can expect more frequent storms, the risk of flooding will increase in London, and Munich will experience more hot days and tropical nights. On the other hand, the fact that the causes of climate change are concentrated in cities also offers an advantage, in that big cities can tackle the problem at its source and climate protection measures will have their biggest impact in such urban centers.

Future City in the Desert

The region that is most determined to become home to the world's first major CO_2-neutral city is in the desert — in Abu Dhabi, of all places, a country that currently has the dubious honor of leading the world in per capita energy consumption. Masdar City will be a metropolis of 50,000 inhabitants. The key here will be energy efficiency. Masdar City will have to make do with only 200 megawatts of installed electrical capacity rather than the 800 MW that cities of a similar size in this climate zone are accustomed to. This will be accomplished with efficient electric devices and lighting systems as well as consistent recycling. Among other things, this will reduce the city's need for freshwater, which would otherwise have to be obtained through an expensive, complex and energy-intensive desalination processes.

According to Sir Norman Foster's London-based architectural firm, which is responsible for key urban planning issues in Masdar City, more than half of the CO_2 reduction will have to be achieved through the design of the city itself and its buildings. Extensive parks will stretch through the eco-city, and cool winds will blow through these fresh air corridors. As in all traditional desert cities, streets will be designed as small alleys rather than broad avenues. Like narrow canals, these alleys will guide wind between the houses, and the arcades along them will provide additional shade. Bearable temperatures will be ensured by cooling systems designed on the basis of Arab wind catchers – also called wind towers – to generate natural ventilation.

Such intelligent architecture is necessary because the cooling of buildings currently accounts for 70 percent of total energy consumption in Abu Dhabi. While the perceived temperature in Abu Dhabi today is more than 70 degrees Celsius in the summer, the new architectural approach will give Masdar City the feeling of "only" 50 degrees. Intelligent building automation systems will reduce energy consumption even further with the help of sensors that recognize when rooms are empty, for example, and then automatically turn down the lights and the air conditioning.

Whatever remaining energy is needed will be provided by renewable sources in photovoltaic, solar thermal, and wind facilities, as well as at plants that use biofuel made from organic waste. Masdar City will also become a model city for smart power grids that help save energy. People will get around town either by walking, or by using electric vehicles as well as electrically operated mass transit, for example automated cabin vehicles. No gasoline-powered cars will be allowed in the city – a true revolution for a country whose wealth was built on oil. If all goes according to plan, Masdar City will become a model for sustainable cities the world over in terms of renewable energy sources, intelligent building systems, and smart networks – all designed and built in cooperation with numerous international firms and renowned institutes. Fittingly, the city whose name, Masdar, means "source" in Arabic will become a source of inspiration and new ideas for the post-petroleum age.

Munich, October 2050. Julian, a games designer, is proud of his multitasking skills, but this looks like a big challenge. Just as he's getting to grips with a new recipe from his Web community — curried fish gratin in coconut milk with a peppercorn crust — the plastic screen of his communications wall rises to reveal his girlfriend telling him that she plans to bring someone extra along for dinner. Then a shriek from behind alerts him to a do-or-die battle between his brother and a 3-D monster in neon red. A sharp fizz from a laser pistol brings the monster to an untimely end. Scarcely audible, by contrast, is the muted beep from the domestic CHP plant, which has registered the weather forecast and is gradually ramping up the heating in readiness for the arrival of a cold snap. Water from the faucet, meanwhile, is still thundering into the sink, though it does switch automatically to cold. That's clear because an LED light is coloring it blue rather than red. And to cap it all, the alarm goes off because the oven door is open. No wonder Julian drops an egg in all the commotion. Nevertheless, his new cleaning robot is as vigilant as ever and, like a dutiful butler, cleans up the mess in seconds. With a sigh of resignation, Julian tells the control system to cut the water, turn off the oven, and bring down the partition that separates the games room, before setting about reassembling the ingredients for a perfect evening meal...

A POWER PLANT IN EVERY BASEMENT, STARS IN EVERY CEILING

For thousands of years human dwellings comprised little more than a roof for shelter against the elements and walls to keep out enemies and predatory animals. Later, buildings acquired a variety of functions and became stores, workshops, fortresses, and palaces. The Romans devised a system of underfloor heating and built aqueducts to transport water. Sewers and toilets soon followed. Industrialization brought electricity and a host of conveniences with it: the first electrically-powered elevator and electric light in the 1880s, the vacuum cleaner around 1905, the radio and the refrigerator in the 1920s, the electric stove, TV, and hair dryers in the 1930s, and the microwave in the 1970s. But a house still remains little more than protective shell, packed with appliances that function independently of one another and the outside world.

By 2050, however, this situation will have radically changed. "Buildings and cities will be equipped with an electronic nervous system and acquire their own intelligence," prophesies William J. Mitchell, a professor of architecture at the Massachusetts Institute of Technology in Boston who heads a research group investigating the city of the future. "Individual houses and residential areas will develop into intelligent organisms that are aware of what's going on inside and around them and will react accordingly." The comparison with an organism may be overstated, but it seems evident that buildings are set to become smarter and more efficient in the future. After all, it doesn't make sense for today's standard automobile in the garage to be far more intelligent, responsive, and communicative than our own home.

Air Conditioning that Knows Tomorrow's Weather

Much progress has been made in reducing the energy consumption of buildings. Modern housing built according to passive house standards — with superinsulated walls, ceilings, and windows along with a ventilation system with heat recovery — has an annual heating requirement of a mere 15 kilowatt hours per square meter. This is less than ten percent of the heating required by buildings constructed before 1984! The figure of 15 kWh/m²a is the limit below which a house, on average, no longer requires any active input of energy for heating — hence the name "passive" building. Most of the energy required for heating is generated by the occupants themselves and the electrical appliances in the house. This is then stored in the floor, walls, and ceilings, which emit the warmth when the temperature falls.

If additional heating is required, heat pumps can be activated. Powered by electricity, these use warmth from the air outside or from the earth to provide heating and hot water. A heat pump can generate as much as four or more kilowatt hours of thermal energy from just one kilowatt hour of electricity. Passive houses are between five and 15 percent more expensive to build than conventional buildings, yet the savings they yield in lower energy consumption pay for the extra cost within years. What's more, in some areas they are eligible for low-interest loans and construction grants. There are already tens of thousands of such buildings in Germany, Austria, and Switzerland, and the trend toward this type of architecture is gathering pace all the time.

The use of smart technology can also cut energy consumption in conventional buildings by 30 to 50 percent. Here, computers match the output of heating, air conditioning, and ventilation to actual requirements and calculate when systems must be activated in order to achieve an ideal indoor climate. In the future this process will even take account of the weather forecast. Then, for example, the home automation computer will be able to turn up the heating when a cold front is on its way and reduce it again at exactly the right time as soon as warmer temperatures are predicted. Sensors are also increasingly being used to detect whether rooms are unoccupied so that the lighting and ventilation can be turned off. In the buildings of the future – initially office and commercial real estate, followed later by housing – building technology will control and coordinate the electricity and water supplies, heating, ventilation, air conditioning, and refrigeration as if it were all a single entity, together with security systems such as fire prevention, intruder prevention, access control, and video surveillance.

Personal Power Generators

By 2050, many houses will not only consume but also generate electricity as active members of a smart grid, the intelligent power supply network of the future (see p. 52). In fact, there are already houses today that are net generators of power, including those in architect Rolf Disch's solar housing development in Freiburg in southwest Germany. In another project that has attracted substantial acclaim, a design by students from Technical University of Darmstadt was awarded first prize in the Solar Decathlon, an international competition for zero-energy buildings in Washington, DC held in 2007. TU Darmstadt also finished ahead of all other projects in the subsequent Solar Decathlon competition, which was held in 2009.

The winning design, which was built with the help of research funds from Germany's Federal Ministry of Transport, Building and Urban Development, is currently making an extended tour of German cities to promote sustainable architecture. For a floor area of around 90 square meters, the building has heating requirements of a mere 12 kWh/m^{2}a. In addition to highly insulated walls, roof, and windows, the building also features special thermal energy storage units based on so-called phase change materials. Photovoltaic cells mounted on the roof and integrated into the sun louvers supply so much electricity that a surplus can be fed into the grid. In addition, solar thermal collectors integrated into the roof generate hot water, and the entire building is largely constructed of renewable, natural, and recyclable materials.

The Fraunhofer Institute for Solar Energy Systems, which is based in Freiburg, Germany, and the BASF company have been largely responsible for advancing the development of phase change materials. Three of the researchers involved in this work were nominated in 2009 for the German Future Prize, also known as the Federal President's Award for Technology and Innovation. Chemist Ekkehard Jahns describes the principle as follows: "The heat accumulators work just like an ice cube. As long as the ice is melting, the cube maintains a temperature of exactly 0° Celsius. Only when the cube has completely melted does the temperature rise. Instead of ice cubes, we use paraffins, wax-like substances that melt at a pleasant room temperature. On warm days they absorb large amounts of heat from their surroundings and thus prevent the ambient temperature from rising any further. At night the wax hardens and emits the stored heat." Volker Wittwer, a physicist and cofounder of the Fraunhofer Institute for Solar Energy Systems, elaborates: "The basic idea behind using such materials is already 50 years old. But the decisive advance was the recent discovery of how to pack these materials into microcapsules."

These minuscule capsules are made of polymethyl methacrylate and have a diameter of five micrometers, less than one tenth the thickness of a human hair. "That yields a number of advantages," Wittwer explains. "The phase transition from solid to liquid takes place within the capsules, so there's no leakage of paraffin. And the large surface area and small volume of all these tiny beads means that a lot of heat can be rapidly stored in, and retrieved from, the material. Furthermore, the capsules can be mass-produced and easily mixed with building materials." They can be incorporated into plaster and mortar for application to walls, or into drywall panels and even wood. The beads are so small that there is absolutely no difference in the appearance of the new building materials, and no problems with drilling holes or inserting nails. A wall containing these microcapsules that is only three centimeters thick offers the same thermal comfort as a concrete wall 40 centimeters thick. This also makes the new materials highly suitable for refurbishing old buildings. In an ideal scenario, they could completely eliminate the need for air conditioning in residential, school, and office buildings.

Because our heating energy requirements are set to be reduced, it will become easier to do away with oil and gas furnaces altogether and convert to smaller systems. These will include not only photovoltaic cells for electricity and solar thermal collectors for hot water but also, and ideally, systems that can generate both power and heat with an efficiency of over 90 percent. In the future, conventional condensing boilers, which can only generate heat, will probably be replaced by micro-cogeneration — or combined heat and power (CHP) — plants. These gas-fired micro-CHP plants will be able to generate both electricity and heat simultaneously. In 2010 a number of manufacturers started to introduce systems of this type with an output of around five kilowatts of thermal energy and up to one kilowatt of electricity.

For consumers, this will mean that they will be able to have a personal power plant in the basement that not only produces heat but also – when in operation – covers two thirds of the power needs of an average four-person household. Another option would be to use fuel cells, which are also powered by a gas. These generate significantly more electricity than a mini-CHP system but are still too expensive. In any event, experts predict a clear trend toward "off-the-shelf" power generation systems for domestic use, whether based on photovoltaic cells, fuel cells, or micro-CHP plants. In other words, it looks like tomorrow's homeowners are set to become small-scale power plant operators!

Windows that Think for Themselves

As with many other next-generation technologies, advances in construction engineering often hinge on the development of new materials. With regard to insulation, such advances include vacuum-insulated panels and aerogels – materials with innumerable nano-sized pores that hinder the flow of heat. As seen above, phase change materials have transformed the nature of energy storage in building materials. Other developments include the use of self-cleaning façade coatings featuring the lotus effect, which causes liquids to pearl into droplets on the surface, thereby removing dirt. And tomorrow's windows could well be equipped with photovoltaic cells made of thin organic plastics. These materials, which are now approaching the production stage, still have lower efficiency and a shorter service life than conventional solar cells made of silicon, yet they are also transparent and therefore ideal for installation between the glass panes of insulated glazing.

Architects also like to use windows made of electrochromic materials, whose opacity can be precisely regulated. This type of glass is already being used in some automobile rearview mirrors in order to reduce headlight glare from a following vehicle. A sensor measures the light intensity and dims the mirror accordingly. Initial projects to investigate applications for windows have now been launched – for example, at the Science College Overbach in Barmen, Germany. Opened in 2009, this facility is housed in a passive building that features a raft of energy-efficient technology, including a special electrochromic coating for its triple-glazed classroom windows, which makes the glass turn blue when a low voltage is applied, thus protecting the computer workstations from sun glare and excessive heat. As soon as the voltage is reversed, the windows become fully transparent.

The Forum of the Science College also features another architectural innovation: heliostats installed in the ceiling. These guide daylight into the building by means of mirrors that automatically track the sun. As ergonomics experts agree, daylight is the best kind of light for people. A blend of daylight and artificial light therefore not only saves energy but also increases well-being. One solution here is to use optical fibers to transport daylight to places where it's needed; another is to install windows fitted with small prisms which in winter, when the sun is low, direct a lot of light into a building, and in summer, when the sun is high, reflect part of it and thus help to keep the interior cool.

Water with a Message

Most experts agree that by 2050 buildings will also feature completely different types of indoor and outdoor lighting. Two technologies in particular are set to revolutionize this field of design. The first is based on the use of dot-like light-emitting diodes (LEDs) that use only one fifth of the energy needed by conventional incandescent light bulbs but have many times their service life. LEDs in all colors, including brilliant white, are now available. In automobiles they have come to replace conventional lamps throughout the vehicle, from display lights to tail lights and headlights.

LEDs are also the favored technology for creating dramatic lighting for architecture. Whereas Europeans prefer white LEDs, the skyscrapers of Dubai, Singapore, and Shanghai shimmer in bright colors, with each building resplendent in its own individual lighting. At last year's World Cup, even the soccer stadium in Durban featured a huge arc aglow with LEDs as an emblem of the new South Africa. During the Advent period in 2009, the manufacturer Osram fitted the turbine blades of a 100-meter-high wind generator in Munich with 9,000 LEDs that created circles, stars, and landscapes in the night sky — for only the amount of power it takes to boil two kettles of water. What's more, the electricity came from the wind generator itself and was therefore completely CO_2-neutral. In the future, more and more artists and advertisers will be using 3-D laser displays and large illuminated video panels to attract the public's attention. Local authorities may have their hands full deciding whether to stop increasingly common light pollution so as to prevent sensory overload.

LEDs are also set to have a creative influence in the home. Today, LED lamps are available in the traditional light bulb shape, but they are so small and emit so little heat that they will be increasingly inserted into tables, cabinets, jewelry, glasses, and even clothing. Indeed, we can soon expect to see designers incorporating LEDs and light fibers into evening wear (see p. 200) — something that has already happened with curtains and carpets. LEDs connected to a temperature sensor could also be integrated into shower heads and water faucets to create situational lighting: When hot, the water jet lights up in a warning red; when cold, it turns blue.

The Sky in Your Bedroom

The second revolution in the field of lighting will be organic LEDs (OLEDs). Unlike the dot-like LEDS, these are large, illuminated surfaces consisting of extremely thin layers of plastic that emit white or colored light when an electrical current flows through them. This homogeneous light is similar to the indirect, diffuse daylight produced when the sky is full of white clouds. In other words, an OLED lighting unit integrated into the ceiling would replicate natural light much better than spotlights made up of lots of small light sources. This kind of lighting could also be made to contain more blue in the morning, which makes people more agile and alert, and incorporate warmer colors in the evening, thus creating a relaxing atmosphere. This is because the human eye contains not only rods and cones but also a so-called melanopsin receptor that only reacts to blue light and therefore strongly influences our sleep-wake rhythm.

OLEDs are roughly as efficient as LEDs. Even with today's technology, which is by no means fully developed, OLEDs emit between two and four

Walls of illuminated plastic. Transparent or brightly shining, according to taste — organic LEDs will create completely new types of lighting experiences.

times as much light as incandescent bulbs and have five times the service life. Furthermore, researchers are confident that they will be able to double or even quadruple this performance over the next few years. Most important of all, OLEDs are both transparent and extremely thin: Their active layer is less than 500 nanometers thick, one hundredth the thickness of a human hair. The first OLED light source, manufactured by Osram, appeared on the market in early 2010. A mere two millimeters thick, this luminescent OLED tile with a surface area of approximately 50 square centimeters emits a warm white light that is five times as bright as that from a typical notebook.

But this is only the beginning, since today's OLEDs are still mounted on sheets of glass, whereas researchers are already working on the development of flexible lighting tiles (see p. 205) and on OLEDs that could serve as both windows and lights. This would make it possible to design transparent light panels or curved room dividers that, at the flick of switch, emit white or colored light and thereby become opaque. Such systems could adapt the lighting mood and color to the desired situation, e.g. a stimulating, meditative, romantic, or working atmosphere.

Designers could also devise new forms of ceiling lighting with OLEDs. In the bedroom, for example, it would be possible to create a night sky with a myriad of stars produced by LEDs, which would change to a rosy dawn as soon as the alarm clock sounds. If we also consider the advent of large wall-sized displays which in the future will be used to show various images or for TV, video telephone, or Internet applications, then it's clear that lighting is set to play a radically different role in our lives.

The only question is whether the use of efficient light sources will reduce the overall consumption of electricity required for lighting — or whether the sheer volume of these new types of lighting will instead offset such gains. The latter scenario seems to be a real danger, since experts forecast that in Asia alone the use of light is set to double by 2020 as people there begin to catch up with the West. Today each American uses, on average, around ten times more artificial light than a person in China and 30 times more than a person in India.

A Sensor that Detects when Air is Stuffy

Whether it's in the fields of power, lighting, security, or health, sensors are set to play a vital role in the buildings of the future. They will measure the movement and intensity of daylight and regulate lighting accordingly.

They will determine temperature and air quality and adjust the heating and ventilation on that basis. They will analyze the freshness of food in the refrigerator; recognize visitors by face, voice, shape or fingerprints; and monitor for gases that might indicate the outbreak of fire or reduce the well-being of occupants. All this will result from advances that researchers are already making in the miniaturization of sensors and the enhancement of computer performance.

Examples here include the gas sensors — for carbon dioxide, for example — developed by a research team led by 49-year-old physicist Maximilian Fleischer at Siemens in Munich. CO_2 is a greenhouse gas that also has a major influence on the indoor climate in buildings. As a result of exhaled air, CO_2 levels in offices full of employees can almost triple to above 1,000 ppm, leaving people feeling tired and unfocused. CO_2 sensors can recognize this situation in advance and indicate that it's time to open a window. But Fleischer's sensors can do even more. His first invention in this field was a small sensor that monitors the exhaust gas from heating systems, thus helping to save energy and cut emissions.

Fleischer, who holds over 160 patents, is currently working on microchips that provide information on air quality — what he refers to as "feelgood" sensors. "In addition to temperature, you need to measure at least three other variables: humidity, gases such as CO_2, and odors," he explains. The surface of his sensors are coated with several layers that react very sensitively to certain molecules. If the sensor detects the presence of such molecules in the air streaming past it, the electrical resistance of the silicon chip is altered and this information is conveyed via a signal.

In other words, the sensors possess a sense of smell, since odors are a mixture of certain molecules. "The organic compounds of certain aldehydes are a good indicator of musty old carpets," he says. If a sensor detects an elevated concentration of these substances in the air, it can instruct an air purifier to release more ozone, which breaks down the odor molecules and thereby removes the smell. Fleischer's odor sensors are also suitable for medical applications. His team has developed a device that analyzes the breath of asthma sufferers for elevated levels of nitrogen monoxide. It has been shown that NO concentration in their exhaled breath rises by three to five times in the hours or even days before an attack. If patients realize this in good time, they can take appropriate action before the inflammation becomes too severe.

Fleischer's team has even developed a sensor that measures breathalcohol levels and is so small that it can be integrated into a car ignition key.

Such a device could tell you if you are still under the legal limit for driving. Also of great interest is the combination of fire detectors with gas sensors. Conventional fire detectors are triggered optically by smoke. "By then, however, anyone close to the source of the fire could have already inhaled toxic gases," Fleischer explains. But sensors can detect the gases typically produced by fires and would react long before smoke has developed.

Some large office buildings and commercial facilities are already fitted with thousands of simple sensors to register temperature, movement, and light. The signals from these are received by a central system that adjusts the air conditioning accordingly, raises or lowers sun louvers, or switches lighting on and off. However, gas and odor sensors are still very rare, and as a rule the technology is too expensive and sophisticated for domestic use. Yet the integration of many sensors into microchips and the advent of cheaper production will radically alter this situation in the coming decades, not least because sensor technology, when used intelligently, can save a lot of money and energy.

Tiny sensors. In the buildings of the future, these microchips will be able to monitor temperature, gases, and odors.

Harvesting Power from the Environment

State-of-the-art sensors don't even need a dedicated power supply, as they can "harvest" the energy they need from their surroundings by means of a photovoltaic cell or a piezoelectric transducer, for example. The latter is based on special crystals that discharge electricity when subjected to pressure. This property, which is exploited by piezoelectric lighters and pressure sensors in cars, can also be induced by oscillation. Piezoelectric devices can therefore generate electricity when they are exposed to an air current or mounted on a component that vibrates. Such sensors can take a measurement once a minute, for example, and still retain enough power to transmit the reading via radio to a central computer or a smart node.

The advantage of such energy-independent sensors is ease of installation, since they require neither wiring for signal transmission nor batteries that need to be replaced — which makes them practically maintenance-free. EnOcean GmbH, a company based near Munich, Germany, has been using batteryless radio sensors for the last ten years. Applications include light switches, which are supplied with power whenever a finger is pressed against them. This short pulse of energy is all that such a microchip requires to turn the light on and off via radio. And since no cables are needed, the switches can be stuck to the wall wherever it's convenient.

Some researchers go even further and predict that in a few decades buildings will feature thousands of tiny energy-independent sensors throughout. Thanks to continuing miniaturization, these so-called microelectromechanical systems (MEMS) — sensors and actuators the size of a grain of sand — will be small enough to be an integral part of carpets and wall paints, where they will act like sensory organs, measuring air temperature, gas levels, and odors before transmitting their readings to a building's automation system. Such systems could also take action themselves, by opening warm air vents, for example, switching the OLED windows to opaque, or activating an ozone generator to neutralize unpleasant odors.

Calling Home

Although the installation of such smart technology will one day be a real option, it seems likely that for most homeowners the cost will remain prohibitive. Nevertheless, the trend is clear. A host of devices will have their own processing and communication capacity. They will be networked, via home control systems, with one another and their surroundings. It will also be possible to control such devices remotely via a secure Internet connection. That classic vacation question "Did I turn the oven off and lock the front door?" will be a thing of the past, since such technology will enable homeowners to check up from faraway destinations and, if necessary, correct such an oversight.

A number of such smart buildings have gone on show in recent years. Back in 2005, for example, practically all the appliances in the "T-Com-Haus" in Berlin, from the TV to the washing machine, were networked and able to communicate. By means of a small mobile device, it was possible to control not only the home appliances but also the lighting, blinds, air conditioning, door, and alarm systems. The system was able to report when the washing machine would finish its cycle and whether the freezer door had been left open. Occupants could jog through virtual landscapes projected onto large screens installed in the fitness room or send and receive video messages on an interactive electronic bulletin board.

One major reason why it is still so difficult to purchase such "smart homes" in all their glory or retrofit homes with such technology is that the many manufacturers in this field — whether of home appliances, multimedia systems or heating, ventilation or air conditioning systems — have found it difficult to agree on common standards for communication networks and control signals. However, an international UPnP (Universal Plug and Play) Forum has now been established to create standards and certify devices that are controllable via an Internet-based network. The use of UPnP interfaces will guarantee that all the appliances and systems of smart homes speak a common language.

Tomorrow's home appliances will also be considerably smarter than today's. Washing machines, for example, will be able to read small RFID labels on clothing so as to determine what kind of garments, and how many, are in the drum. That way, the machine can immediately recognize a colored sock in a load of white laundry and alert the user. If the colors all match, the machine automatically selects a suitable program and the right amounts of water and detergent. Such a machine could also check the current electricity price and, if the owner so wishes, postpone the cycle until a time — late at night, for example — when power is cheaper.

Last but not least, the washing machine of the future will be able to check that it is functioning properly. If such readings depart from nominal values, the machine will issue a fault report or even contact customer support automatically via the Internet. This is already standard practice for large items of medical equipment and for systems in power plants and major office complexes, where any irregularities are reported to remote control centers before the systems actually go down. That way, service professionals can run an online check and sometimes even remedy the glitch during ongoing operation, especially if it is a software problem. And if an on-site technician is required, he or she can wear a helmet camera so that the specialists from customer support can analyze the images from a distance and provide precise instructions as to what to do.

Personal Butler

Also of great interest is the extent to which robots will become a permanent fixture in the households of the future. The T-Com-Haus of 2005 already featured small automatic vacuum cleaners that tirelessly cleaned the rooms, elegantly avoiding any obstacles in their path and returning to a charging station whenever they ran out of power. According to the International Federation of Robotics, there are now over ten million service robots in action worldwide, mainly in Japan and the U.S. Household robots account for around half of these, including robots to clean floors, windows, and swimming pools; surveillance robots equipped with webcam, microphone, and loudspeaker; and robots to mow the lawn.

The other half comprises a whole variety of robotic toys, mainly in the form of cuddly animals ranging from seals to dinosaurs. Very popular in Japan, for example, is a white robotic seal named Paro that reacts to touch, light, temperature, and sound. It can also recognize its name or a spoken greeting and can reply accordingly. It can even learn and behave in ways people like — for example, repeating a certain action in order to be stroked.

Gentle helper. A robot developed in the DESIRE project is able to identify specific objects and carefully pick them up.

Paro is already being used in Japan for therapeutic purposes, as it is ideal for calming patients with dementia and reducing their stress levels.

With the world's population aging and people still wanting to live independently for as long as they can, most experts believe that robotic assistants will be increasingly required and accepted in the home, where they can lift heavy loads, clean, perform tasks in the kitchen, or read out loud. Robotic wheelchairs for handicapped people have been developed. Equipped with cameras and arms, they can reach for food and beverages. In the DESIRE (Deutsche ServiceRobotik Initiative) project, participating organizations have built a robot that can recognize and safely pick up around 100 different everyday objects, even if they are partially concealed by other objects.

Similar projects include a robot built by researchers at the University of Karlsruhe in Germany that can load and empty a dishwasher, and Care-0-bot, developed by the Fraunhofer Institute for Manufacturing Engineering and Automation in Stuttgart, which is designed to be an electronic butler.

The latest version of Care-0-bot can recognize typical household objects and safely pick them up with its three-finger hand — even fragile glasses — place them on a tray, and navigate to its destination without colliding with anything. With the help of 3-D cameras, laser scanners, and a map of its immediate surroundings, it can orient itself, open doors, avoid obstacles, and move securely among people. In a trial at a retirement home in Stuttgart, Germany, the robot made a polite bow after it had brought drinks for residents.

Read all about it: Robots Beat Soccer World Champions in 2050

It would not be difficult to install advanced voice-control and gesture-recognition technology in a robot. In principle, even today a suitably designed robot can understand commands combined with a pointing gesture, such as "Go there and get me a glass" or "Put the plate in the dishwasher over there." Some robots can even correctly interpret facial expressions and emotions in the voice. There are also robots that can climb stairs or play soccer. Indeed, the organizers of the annual international RoboCup, which was inaugurated in 1997, have set themselves an ambitious target — that a team of robots should beat the reigning soccer world champions in a game to take place in 2050!

Even though it seems inconceivable that soccer-playing robots would be able to win against Brazil, Spain or Italy, it is not unlikely that electrical goods stores of the future, perhaps within 15 to 20 years, will have cleared the shelves of PCs to make way for PRs — Personal Robots. It seems probable that the first robotic assistants will be found in factories, hospitals, and nursing homes, before gradually capturing the market for private customers.

Equipped with intelligent software, such personal robots would soon learn their owner's preferences and take account of them in their behavior (see p. 193). (see p. 193) After a certain training period, no two PRs anywhere in the world would be the same, even though they were originally supplied by the manufacturer with exactly the same functions. The obstacles in the way of such a vision are less of a technical challenge than an economic, social, and legal question. For example, will such complex systems be relatively cheap to produce? After all, even for well-off seniors they shouldn't cost more than a compact automobile. Will people accept robots as fellow residents of their own homes? How humanoid should they be? And who will be legally responsible if an autonomously acting personal electronic butler of the future blunders in some way or even accidentally injures a human being?

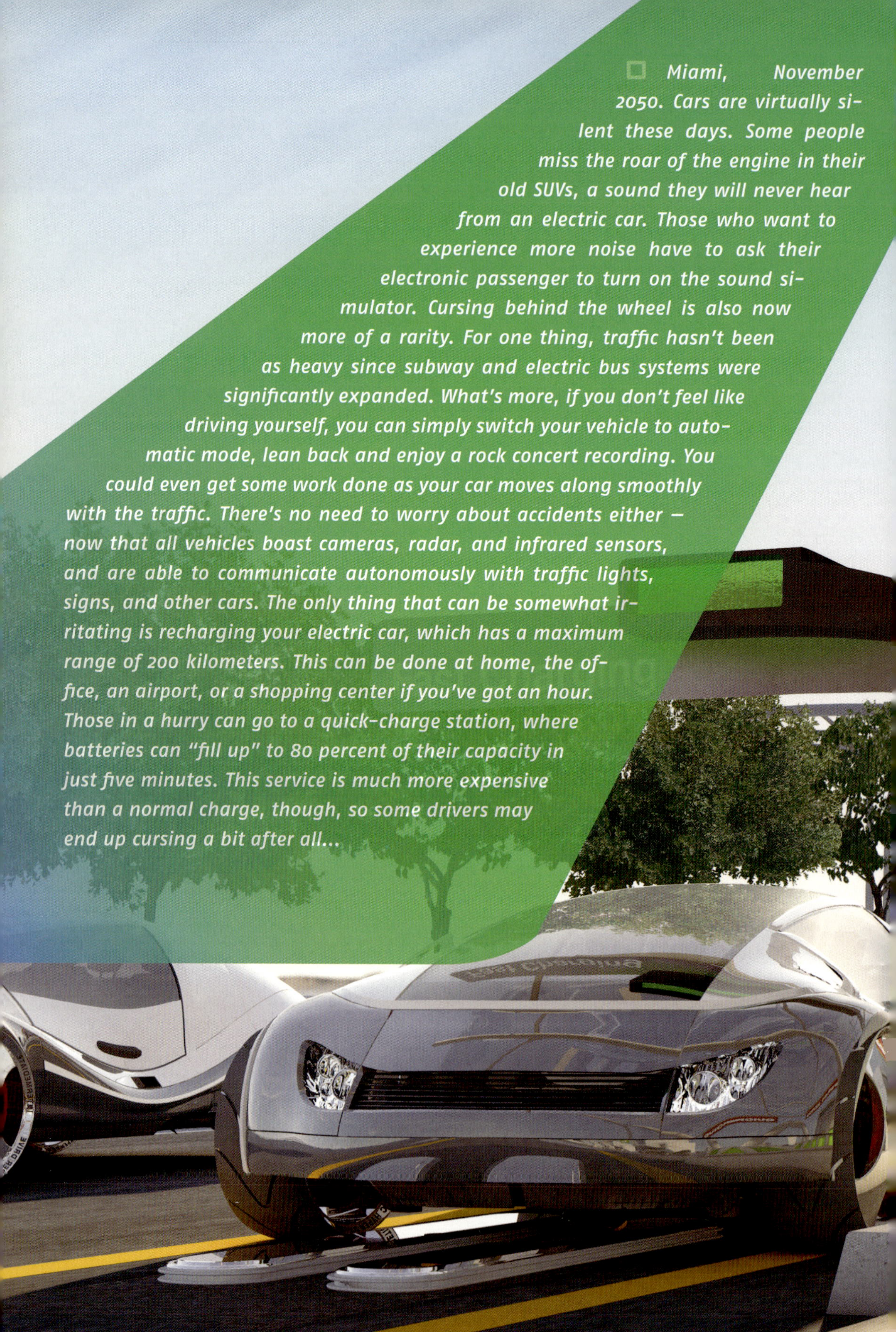

Miami, November 2050. Cars are virtually silent these days. Some people miss the roar of the engine in their old SUVs, a sound they will never hear from an electric car. Those who want to experience more noise have to ask their electronic passenger to turn on the sound simulator. Cursing behind the wheel is also now more of a rarity. For one thing, traffic hasn't been as heavy since subway and electric bus systems were significantly expanded. What's more, if you don't feel like driving yourself, you can simply switch your vehicle to automatic mode, lean back and enjoy a rock concert recording. You could even get some work done as your car moves along smoothly with the traffic. There's no need to worry about accidents either — now that all vehicles boast cameras, radar, and infrared sensors, and are able to communicate autonomously with traffic lights, signs, and other cars. The only thing that can be somewhat irritating is recharging your electric car, which has a maximum range of 200 kilometers. This can be done at home, the office, an airport, or a shopping center if you've got an hour. Those in a hurry can go to a quick-charge station, where batteries can "fill up" to 80 percent of their capacity in just five minutes. This service is much more expensive than a normal charge, though, so some drivers may end up cursing a bit after all...

ELECTRICITY: THE FUTURE OF DRIVING

"Back to the Future" would be a great motto — if not a little ironic and exaggerated — for what automotive development engineers are now doing to ensure they can continue putting a successful product on the roads of tomorrow. Basically, researchers are trying to get their metal boxes to do what their flesh–and–blood counterparts have been doing for a very long time. After all, horses avoid collisions, communicate with each other while moving rapidly, keep a close eye on what's ahead and behind, evade obstacles, and listen to the commands of their "drivers." Camels, for their part, are very frugal energy consumers, and the waste they produce is biodegradable. A mule can find its way home — even if its driver has fallen asleep. The vehicles of the future will be able to do all these things and more. With various "sensory organs" they will probe the surrounding environment, establish contact with other vehicles, act autonomously in an emergency in order to avoid accidents, and use their fuels as sparingly and ecologically as possible.

Even today's cars are already more like elegant computers on wheels than crude boxes of steel and plastic. Whereas software accounted for only four percent of the value of a passenger car in 2000, that figure has now reached 15 percent. If you add computer intelligence and electronics, you get about 40 percent of the current cost of a vehicle. Tasks that used to be handled by mechanical and hydraulic systems are now increasingly being carried out by microprocessors. Steering columns are giving way to electric controls — the first-ever joystick-operated vehicles are being put through their paces — and hydraulic brakes are being replaced by electrical brake systems. Specialists refer to this trend as "drive-by-wire" — a development from the aviation industry that is now making its way into automotive technology systems.

All these things are being driven by customer demands for more comfort and safety. The frequently aired complaint that electronic systems constantly fail isn't borne out by statistics, which show that automobiles are more reliable today than they were 35 years ago. According to the German automobile club ADAC, the probability of a breakdown with a new vehicle within its first six years of use has fallen from 3.5 percent in 1976 to 0.6 percent today. However, it's also true that while most breakdowns in the past resulted from leaky lines or other mechanical defects, the main cause today is to be found in electronic system faults. In order to prevent these, and also reduce development costs, leading automotive and electronics companies are now working together to establish standards for hardware and software modules. Success here will also solve another fundamental problem. At present, the average age of cars on the road is around eight years. However, software is developing much more quickly and should therefore be updated frequently. In other words, older vehicles need to be able to take on such updates as well.

Filling up on Electricity and Information

The smart cars of tomorrow will thus fill up not just on fuel but also on information as they continually remain online. They will, for example, be able to access their driver's appointments calendar, and thus be aware of his or her travel plans. If they use the Internet to obtain information on current traffic conditions, they will not only be able to feed this data to the navigation system but also calculate how much fuel is needed for a given trip. Such information will be important to the electric cars of the future because it will enable them to know just how much electricity they need, or even when they can return surplus power to the grid and make their drivers some extra money.

The smart car will be perfectly prepared for its driver when he or she gets into it in the morning. It will recognize its driver, unlock doors, automatically adjust the driver's seat to the right position, and release the ignition lock. It may even issue a greeting via a virtual passenger in the form of an avatar that appears on an OLED-foil cockpit display (see p. 205), which will take up the entire space where the dashboard used to be. This electronic servant, which will also be linked to the "smart home," will then ask drivers if they want to tune into the radio station they were listening to in the kitchen a few moments previously, remind them of important appointments that day, and tell them not to forget to turn on the house alarm system. The vehicle will also be able to collect information on the move — for example, by calling up interesting sites or events in the towns and cities it drives through.

Drivers will speak with their virtual passenger the same way they would with a real person. Such voice command systems can already understand and carry out some 10,000 simple commands, but in the future they will also be able to correctly interpret fluid sentences such as, "Please read me my e-mails from this morning," and even fragmented orders like "Latest song from Shakira." These communication interfaces to the user — via displays, avatars, or voice commands — are vitally important to automakers because they determine if the associated systems will be accepted by customers. Whether or not drivers like their cars depends mainly on what they see, hear, and feel; it does not result from invisible high-tech details under the hood.

Another innovation from the aircraft industry can also now be found in some vehicle models: the head-up display — a mirror system that projects warnings and navigation tips onto the windshield. From the driver's position, it appears as if the information were floating above the hood. As a result, he or she doesn't need to look away from the road to read the data — a feature that enhances safety. Head-up displays might also be equipped with augmented-reality technology in the future. This technology supplements the driver's actual view of the road ahead with virtual, computer-generated information. Specifically, it projects the route to be taken directly onto the windshield as a thick colored line that looks to the driver as if it were literally painted on the road. The driver can thus react intuitively and rapidly to any situation.

Cars that Leave "Odor Trails"

Today's vehicles already "know" when a skid commences, when sharp braking is called for, and when tire pressure drops. These events are registered by anti-lock braking systems, electronic stability control units, and tire pressure monitors, for example. But in the future, sensors could also send this data on to vehicles behind, which would then react instantaneously to the information. The same applies to information on a traffic jam that has formed after a curve or on the other side of a hill; the ability of vehicles to notify one another of such a hazard would prevent many accidents. All they would need for this are radio transceivers with a range of just a few hundred meters.

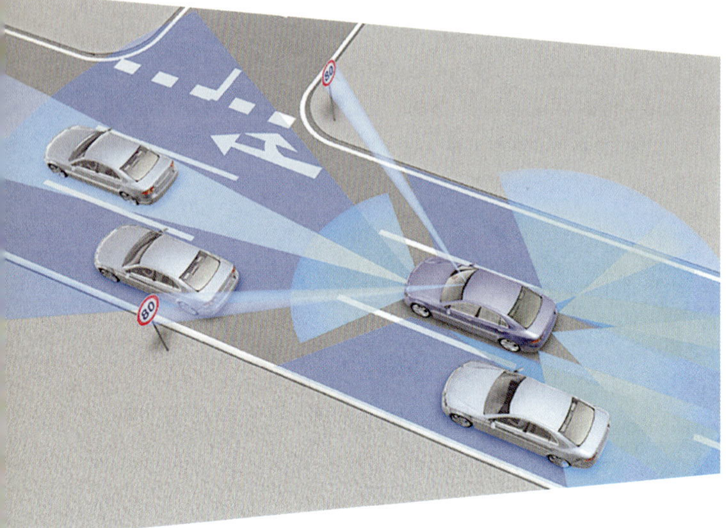

Networked cars and infra-structures. The automobiles of tomorrow will recognize traffic signs and watch what other vehicles are doing.

Alternatively, cars could do what ants do: leave "odor trails" — or digital pheromones, as researchers call them. Here, a vehicle collects relevant information for each route segment, such as the amount of time needed to complete it. The vehicle then transmits the data anonymously to a computer, which in turn forwards it to other vehicles in the vicinity. Each car thus has an overview of the actual traffic situation and can therefore select the optimal route. Data on the current flow of traffic can even be compared to previously-recorded information to generate traffic jam forecasts. The better the data, the more favorable the environmental impact as well. In Germany alone, for example, some 20 percent of all fuel burned is wasted in jams and slow moving traffic. Eliminating such congestion would save around 12 billion liters of fuel per year and reduce CO_2 emissions by 30 million tons.

The European Union set itself an ambitious safety target in 2001, when it announced plans to halve the number of traffic fatalities on European roads from nearly 50,000 to 25,000 by 2010. Most EU members were unable to reach their part of this target, despite some major successes in countries such as Germany, where fatalities in traffic accidents were reduced to approximately 4,000 in 2009, down from 7,000 in 2001. However, a look at statistics from 1970 shows the true extent of the advances that have been made here. In 1970, there were 19,2000 traffic fatalities in what was then West Germany alone, or nearly five times as many deaths as in 2009. What's more, the statistics for 2009 also include the former GDR. This sharp decline is largely due to the introduction of seatbelts, airbags, and anti-lock systems. Still, achieving the EU's 50 percent reduction target, not to mention the even more ambitious vision of "accident-free driving," will require more than these purely passive systems. The key is active safety — in other words, cars that "see" and prevent accidents from even happening in the first place.

Vehicle Eyes and Ears

The cars of the future will therefore be equipped with a variety of sensors that monitor the traffic situation, warn drivers in emergency situations, and even intervene when it becomes clear that a driver can not react in time. Initial versions of such sensors are already being used. Some employ radar and infrared lasers that monitor the vehicle's surroundings in order to prevent rear-end collisions, while others make blind spots visible or warn drivers not to pass when a car is approaching too fast from behind. There are also ultrasound sensors that help with parking, thermal imaging cameras for improving nighttime visibility, and video cameras that can monitor the road ahead and recognize traffic signs.

Also very promising is the use of a combination of different systems. For example, light-emitting diodes in headlights could be controlled so quickly and precisely in the future that they would shine directly on a darkly clothed pedestrian identified by a night vision system. Video cameras and infrared lasers integrated into windshields could also work in unison with radar sensors behind bumpers. An onboard computer can use data from the video camera that follows lane markings to automatically keep a vehicle in its lane — even in curves. In combination with a laser sensor, a radar system can measure the distance to, and the speed of, vehicles in front. Such an assistance system could automatically drive a vehicle at relatively low speeds — a feature that would enable the car to follow the vehicle in front in slow-moving traffic. In the fall of 2010, a fully automated vehicle was tested on German roads for the first time ever. The "Leonie" research car — a modified VW Passat — had no problem traveling around the city of Braunschweig, even at speeds as high as 60 kilometers per hour. Meanwhile, in California, seven test cars drove 1,000 miles without human intervention in late 2010.

Experts predict a bright future for driver assistance systems, especially since there will be more and more elderly people on the road who will want to remain mobile. It's quite conceivable that old – and perhaps even some younger – drivers will prefer the use of an automatic vehicle mode that enables their car to drive itself, or even allows them to tell the vehicle to park itself after they get out.

It will be important here not to market such vehicles as being for seniors, but as being for anyone interested in safety. An automobile must be developed for all drivers aged between 20 and 90, while still focusing on the needs of the older generation. "Design for the elderly and you'll include the young; design for the young and you'll exclude the elderly" is the rule of thumb for everything from cell phones to computers and cars.

Sensors will not only be used to monitor the traffic environment, but also to check up on vehicle interiors and even the driver. In the future, for example, carbon dioxide will increasingly replace gas containing fluorine as a coolant, as the latter damages the ozone layer. But low levels of CO_2 in a car can cause drivers to get tired. The feeling is similar to that experienced in a poorly ventilated vehicle full of passengers. In such a case, the critical level of 1,000 ppm CO_2 could be reached in just 15 minutes. In the future, gas sensors will sound an alarm should such a situation arise. Drowsiness warning systems could also help to greatly improve safety. Studies have shown that 25 percent of all accidents on German highways are caused by overtired drivers. Systems capable of measuring steering movements and other data that indicate a driver may be on the verge of nodding off are already available today. In the future, it may also be possible to use a small camera in order to determine a driver's stress level. The smart car could then recommend a break or even reduce stress by turning down the radio or eliminating all but the most important cockpit displays.

Many assistance systems are already technically feasible today – but whatever the system, the important thing is that the driver should always have ultimate control over the vehicle. Liability considerations alone make it imperative that the role of electronic helpers be limited to support functions. Assistance systems should only intervene in those instances where a driver is no longer able to control the vehicle. This is already the case with units that recognize emergency situations and automatically provide the required braking pressure to bring the car to a halt, for example.

End of the Oil Age

While it remains to be seen just how much help drivers in the future will want or accept from their favorite toy, one trend will never disappear. Owning one's own vehicle will always be viewed as the fulfillment of the dream of individual mobility. There are now more than 700 million passenger cars on the road worldwide, and experts from Shell predict that this number will double by 2030 — and perhaps even triple or quadruple by 2050. Cheap cars are all the rage in China and India, and oil consumption is rising tremendously as a result. Already today, nearly two billion tons of crude oil is refined into gasoline or diesel fuel for motor vehicles each year. A critical dilemma is approaching, however, because most experts predict that oil production will peak before 2030. In other words, twice as many vehicles will be on the road at a time when petroleum is becoming scarcer and more expensive. Such a situation can not be sustained — even if you leave out the negative impact it will have on the environment. The only realistic solution is to significantly reduce the fuel consumption of each vehicle, increase the use of natural gas and CO_2-neutral biofuels (see p. 56), and initiate a transition to electric cars.

There's no doubt that much has already been achieved in terms of fuel economy. Whereas a BMW 3 Series guzzled 12.5 liters of gasoline per 100 kilometers on average in 1975, today's model consumes around seven liters, even though its weight and engine output are both 50 percent higher. The same is true of the VW Golf, which consumed ten liters of fuel per 100 kilometers in 1975 but only needs around six liters today. It's conceivable that the efficiency of gasoline engines can be raised by an additional one-third between now and 2020. Technically speaking, the key thing is to achieve the most uniform combustion possible at homogenous and relatively low temperatures in the combustion chamber, which, in turn, minimizes nitrogen oxide emissions. Combined diesel–gasoline engines or customized synthetic fuels have a role to play here.

However, this is not what executives at the world's major automakers have in mind when they talk about "reinventing the automobile." Instead, the revolution they're planning revolves around electric vehicles. Back in 2007, it became clear that a new attitude was developing — in Hollywood of all places. That was when it suddenly became "uncool" to drive around in large limousines or flashy sports cars. Leonardo di Caprio was one of the first to drive a low-emission Toyota Prius hybrid with a combined combustion engine–electric motor drive system. Julia Roberts and Cameron Diaz followed, and George Clooney even went completely electric. Hollywood thus went green.

Screenplays are now printed on paper made out of hemp, while film studios have installed solar panels on their roofs, switched over to hybrids or electric cars, and started making CO_2-neutral movies — the first of which, fittingly enough, was Roland Emmerich's 2004 climate disaster film *The Day After Tomorrow*. Having calculated that producing the film had generated 10,000 tons of CO_2 emissions, the studio responsible paid compensation in the form of a $200,000 investment in reforestation projects. However, it wasn't until former U.S. Vice President Al Gore won two Oscars in March 2007 for his climate change documentary *An Inconvenient Truth* that the paradigm shift in the film industry finally entered the mainstream.

Business consultants estimate that one out of every four new vehicles in Europe will be either a hybrid or electric model by 2020. At that time, there might also be 20 million purely electric cars worldwide. Perhaps by 2040 or 2050, most vehicles on the road will be powered by electric drive systems — a fact that doesn't automatically translate into excessive power consumption. If all the passenger cars in Germany today were to be powered by electric motors, annual mean electricity consumption would rise by only 16 percent, or by approximately 0.3 percent per million electric vehicles. At the same time, consumption of gasoline and diesel fuel would decline by 30 million tons per year.

From Hybrids to Pure Electric Drives

The first step in this development will involve the use of purely electric vehicles for urban transport and as second cars, neither of which will be driven more than 70 kilometers per day — not even on weekends. That would add up to nine million passenger cars in Germany alone. Vehicles driven for longer distances will continue to require an electric motor and an additional combustion engine for some time, however. These cars could take the form of a classic parallel hybrid in which both propulsion units power the vehicle. Here, the batteries and electric motors would mainly go into action over short distances and use recovered braking energy to accelerate the vehicle again.

However, another alternative will increasingly come to the fore: the more efficient series hybrid. Here, the vehicle is driven exclusively by an electric motor while a small combustion engine produces electricity for the battery via a generator when the battery charge falls below a certain level. Unlike a pure electric car, such a vehicle would not have a limited range. What's more, the combustion engine — whether powered by gasoline, diesel, ethanol, or biofuel — would always run in an efficient operating mode and therefore emit low levels of CO_2.

Many buses are already equipped with such a series hybrid system. The latter is also being used in the new Opel Ampera — a compact car with a range of approximately 60 kilometers in the pure electric mode. Thanks to its gasoline engine, the vehicle can manage more than 500 kilometers on a full tank. In principle, it would also be possible to replace the combustion engine in a series hybrid with a fuel cell that provides the electric motor with electricity produced from hydrogen or methanol without the need for combustion. When it comes to converting the energy from the fuel they use into electricity, fuel cells are at least twice as efficient as a combustion engine-generator combination. However, because they contain costly membranes and platinum catalysts, fuel cells will probably remain too expensive for the mass market in the foreseeable future.

No More Expensive than Gasoline — but a lot more Environmentally Friendly

The Opel Ampera is equipped with a lithium-ion battery that weighs 175 kilograms, has an energy content of 16 kilowatt-hours (kWh), and takes up just a little more space than a suitcase. However, such batteries currently account for 30 percent to 40 percent of the total cost of an electric vehicle, which translates into € 700 to € 1000 per kWh of storage capacity. This high cost is mainly due to the fact that so far no one has set up an automated manufacturing process for mass producing these batteries. This is about to change, however, because demand for the power packs is rising rapidly.

Experts believe that energy storage costs will decline to € 300 per kilowatt-hour by 2025, at which time a 16-kWh battery would cost less than € 5,000. A 100-kilometer trip would then generate costs of around three euros for the electricity and four euros for wear-and-tear on a battery whose storage capacity is expected to decrease by 20 percent after around 1,000 charging cycles. In other words, the costs involved would become comparable to those of an economical combustion engine — and that's without taking into account the anticipated price increase for gasoline, or additional costs for CO_2 emissions.

In any case, nothing can top an electric car when it comes to climate protection. Even with today's electricity mix and its high proportion of fossil fuel sources, an efficient electric vehicle is already as climate-friendly as a car powered by a combustion engine that consumes only three liters of fuel per 100 kilometers. In addition, the more electricity is produced from renewable sources in the future, the better the climate balance will be. Nevertheless, it's not environmental protection at all that's driving automakers to focus on electric cars but instead sharply rising oil prices and the knowledge that petroleum reserves are running out.

There's good reason why the motto for electric cars is also "Back to the Future." Electric drive systems are actually older than the combustion-engine car invented by Carl Benz in 1885. Four years prior to Benz' innovation, Gustave Trouvé presented a three-wheeled automobile equipped with a lead rechargeable battery at the International Exhibition of Electricity in Paris. In 1882, Werner von Siemens demonstrated to an amazed audience an electrically operated carriage known as the Elektromote that ran on a 540-meter test track near Berlin. The year 1905 saw the launch of the Elektrische Viktoria, the first electric car to be built in small-batch production. It was operated as an elegant hotel taxi, delivery vehicle, and small bus in Berlin at a time when the only vehicles on the road in the German capital besides the first electric streetcars were horse-driven carriages. Later Viktoria models even featured brake energy recuperation technology.

It wasn't until the second decade of the 20th century that electric vehicles were eclipsed by automobiles with combustion engines. Now — one hundred years later — they're back and on the verge of a final breakthrough. There will be losers today, too, however. The future looks less bright, for example, for automotive suppliers that manufacture exhaust systems, catalytic converters, transmissions, crankshafts, and gasoline tanks — none of which are required in electric cars. These companies will need to refocus their activities in order to stay in business. The new electric vehicles are actually even more dynamic than those equipped with combustion engines because their electric motors deliver full torque right from the start. Indeed, electric sports cars that accelerate from 0 to 100 kilometers per hour in less than five seconds already exist.

New Players Join the Fray

A new pioneering spirit is sweeping the automotive industry. BMW plans to manufacture the body for its new "Megacity Vehicle" with ultra-light carbon fibers in order to reduce weight and make long trips possible despite the model's heavy battery. More and more automakers are now also partnering with energy companies to conduct tests on both electric vehicles and interfaces with the power grid. Daimler, BMW, and VW are cooperating, for example, with energy suppliers RWE, E.ON, Vattenfall, and EnBW, while Siemens is involved with all aspects of the new electric mobility technologies. On the vehicle front, Siemens produces electric motors, charging systems, and electronic components, among other things. In addition, it is also developing systems for vehicle-to-grid communication, energy distribution, electricity metering, and power connections to buildings and smart grids. One of the things researchers are working on is fast-charging technology with 120-kilo-watt output, which would make it possible to charge electric vehicle batteries in just a few minutes. They are also developing concepts that would allow many cars to recharge at the same time without overloading the grid. After all, thousands of vehicles simultaneously tapping into the network at a stadium or airport would require the output of a medium-sized power plant.

The German government is supporting test operations in many model regions, while other countries are assisting consumers with electric vehicle

Fast charging at the electricity filling station. No one wants to wait for hours to fill up when they're on the road, which is why researchers are testing methods for charging electric cars in minutes.

purchases. For example, the U.S., Japan, China, France, and the UK pay cash or issue tax vouchers that often total € 6,000 or more per vehicle. But the real revolution is mainly occurring in Asia, where completely new players are entering the field. Build Your Dreams (BYD) is one such company. Headquartered in Shenzhen, China, this firm has been manufacturing rechargeable batteries for cell phones and laptops since 1995 and now accounts for around half of all mobile phone battery sales around the world. All in all, companies from China, Japan, and South Korea cover around 90 percent of the global market for lithium-ion batteries.

BYD is looking to do more, however. That's why the company, which has roughly 60,000 employees, has now launched automobile production operations. Its goal is to manufacture very inexpensive hybrid and electric vehicles in order to gain new customers in Asia, Africa, and South America – and later in the U.S. and Europe. Future market potential is huge in China especially, where the automobile boom has only just begun. Every year, more than 20 million Chinese citizens purchase electric bicycles and motorcycles. And why shouldn't China's story be a re-run of Europe's in the 1950s and 1960s? First, consumers purchase small two-wheeled vehicles; later they switch to affordable cars as prosperity increases.

Western firms have little hope of winning a price war with Chinese companies. Their only chance is to "offset the price disadvantage with a quality bonus," as one executive once put it. Western companies can only gain advantages through their innovative technical solutions, creative designs, and decades of experience in building safe and comfortable automobiles. They need to take seriously their claim of "reinventing the automobile" and completely redesign the entire system architecture using new electronics, sensors, and software.

Something similar happened with the mechanical typewriter. The decisive breakthrough wasn't due to a half mechanical, half electrical device that was able to display two lines of text on a tiny screen. In fact, the true revolution didn't even tale place within the typewriter sector, which stubbornly stuck to its old ways and only made minor modifications. Instead, the great transformation was triggered by personal computers and word processing programs that were able to show a complete text onscreen and print the file.

Why Motors Will Migrate to the Wheels

One of the most important changes that will come with electric cars will be the elimination of the entire powertrain, as tomorrow's motors will be mounted directly on the vehicle's wheels. This concept is also as old as the automobile itself. Back in 1900, Ferdinand Porsche designed an electric car equipped with such motors for presentation at the Paris World's Fair. The concept will become far more sophisticated in the future, however, thanks to drive-by-wire systems. In addition to the drive system, the latter will integrate the steering, shock absorption, and braking systems into the wheels. The art of automobile design will then mainly involve implementing intelligent electronic controls and continually improving energy efficiency. An electric motor is already three to four times more efficient than a combustion engine — and if you mount it near a wheel you eliminate the need for several components, such as heavy, bulky differentials, which generate efficiency losses.

Such motors will also make completely new vehicle architectures possible, since the elimination of a central engine, transmission, and axle shafts substantially increases the amount of available vehicle space. Moreover, control units won't have to be placed directly next to an electric motor, which means designers will have new opportunities to design components such as wheels that can be driven independently of one another. This will not only enable cars to react more effectively to dangerous situations; it will make parking easier, for example. What's more, vehicles thus designed wouldn't need a center console or rigid steering columns. They could be operated with joysticks and fitted with fold-out seats — a feature that would make entry and exiting easier for seniors.

Vanishing Traffic Signs?

But it's not just cars themselves that will change; the entire concept of mobility could also undergo extensive transformation. Inspired by the motto of the World Economic Forum in Davos — "How do you make the world a better place?" — Shai Agassi, a former member of the Board of Management of computer software firm SAP, founded a company called "Better Place" in 2005. Agassi would like to make cars available in the same way that mobile phones are offered today — in other words, subsidized (low-price phone with a contract), or even complimentary. Customers would then not actually buy a vehicle but instead purchase "range," whereby their bill would be based on the number of kilometers driven.

Alternatively, customers could buy a car — but its battery would be owned by Better Place. Agassi's idea here is that so-called mobility kilometers can be offered at a more reasonable rate with electric vehicles than is possible with combustion-engine cars. Discharged batteries could either be recharged in minutes at fast-charging stations or turned in for new ones at

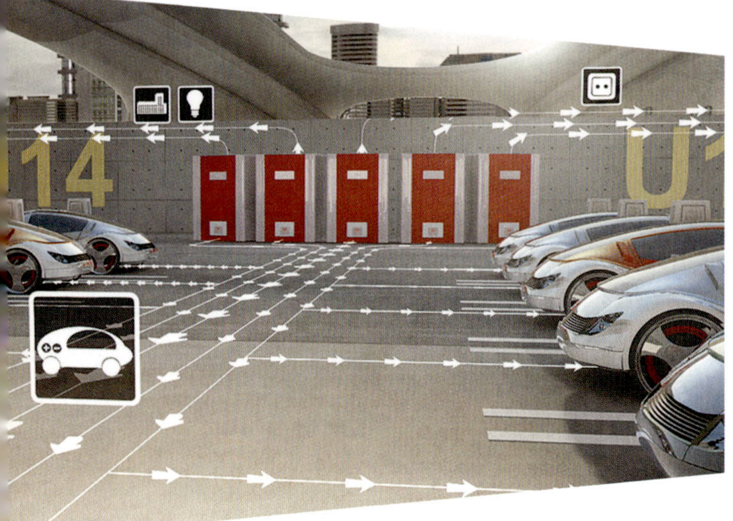

Intelligence everywhere. In the future, electric cars will charge up and feed power back into the grid — whether at home, in a parking lot, or at an airport.

robot-operated battery exchange stations. Agassi has already gained Renault-Nissan as a partner to build his electric cars, and Better Place plans to launch its first pilot projects in southeast Australia, on Hawaii, and in several countries, including Israel, Denmark, and Japan.

The concept of mobility will likely be viewed much differently in 2050 than it is today — at least in the wealthy nations, where green urban districts will offer extensive possibilities to move around on foot or by bicycle. And in much the same way as many cities now operate bike rental stations, the future will bring city-car rental systems that will require only one-time registration and offer easy payment via a chip card or cell phone. Tele-working will eliminate many commutes, and a large number of urban residents will prefer to switch over to significantly expanded public transport networks, especially since the operation of vehicles with combustion engines will be subject to CO_2 fees and city tolls, making it very expensive.

Distribution centers will also be set up on the outskirts of cities. Here, merchandise will be loaded onto low-noise and emission-free commercial vehicles that will be allowed to enter city centers. Buses and trains will not only be very comfortable and energy efficient; they will also be largely automated and run at very short intervals. As a result, more people will use them. Those who nevertheless choose to drive their own vehicle will in most cases use an electric car that can be recharged anywhere electricity is available, whether at home, at a supermarket, the company parking lot, or an airport.

A variety of vehicle sensors will make such a vehicle something of a personal robot on four wheels connected to its environment via numerous networks. Traffic, weather data, and information on places nearby — from parking spaces to restaurants — will all be brought into the car via the Internet. This smart car will also use encrypted data connections to remain in contact with its owner's smart home. The new Galileo satellite system already provides vehicle positioning information down to the last meter, and smart cars could also use radio to maintain connections with other vehicles and the transport infrastructure.

Contact could also be established with "intelligent traffic lights" that remain green for as long as intersections are free. Then, a few more years down the road, when all vehicles are equipped with such systems, it may become possible to abolish traffic signs and lights. In this scenario, a vehicle approaching an intersection where a stop sign once stood would project a warning for its driver onto the windshield — or do nothing if the sensors detect that no pedestrians or other cars are in the vicinity. Now that would be truly intelligent driving.

□ Singapore, December 2050. When the chef at the gourmet restaurant in the Green Tower wants fresh fruit, vegetables untainted by pesticides, or a succulent saddle of lamb, he doesn't have to go far. It's just a short trip with the elevator to the organic farm on the roof of the skyscraper, where there are bushes and trees laden with mangos, bananas and other tropical fruits. Small agricultural robots at the farm monitor free-range chickens and sheep. Even exotic butterflies evidently feel quite comfortable here. There are hidden sensors everywhere to monitor nutrient levels, moisture, and temperature. Solar cells and wind turbines provide electricity. Computers control light levels, ventilation, and irrigation in a perfectly organized ecosystem. If a patron wants to know exactly where the food comes from, the chef guides him up here personally and explains the workings of an organic farm, which minimizes transportation costs and eradicates the need for pesticides. So far, the visitors have all been impressed — and not just by the unique view of Singapore. From up here, it becomes apparent that the city-state is dotted with hundreds of small parks and green terraces on buildings. Although it was a developing country as recently as the 1960s, Singapore has become one of the winners in the age of globalization. As a result, it has meanwhile become one of the most pleasant cities to live in.

FARM IN A SKYSCRAPER

Fruit and vegetable gardens complete with livestock farming in a skyscraper? To some, this scenario may sound like a far-fetched vision of the future, but Singapore is striving to make it a reality. "We've launched a program that supports the planting of green areas on building rooftops," says Richard Hoo from the Urban Redevelopment Authority. "We want to plant 50 hectares in and on buildings by 2030. To do so, we'll make use of rooftops, facades and terraces." The hanging gardens of Singapore will also serve as a natural air-conditioning system and lower the temperature in the city by several degrees. Transforming itself into a city of gardens has become a mission for the city-state, where five million people are crowded together in an area smaller than Beijing's downtown. A large number of exotic plants are already thriving in the narrow spaces between tall buildings, and there is lush rainforest growing just a few miles from the city center.

The vision of "vertical farms" is being pursued by researchers like Dickson Despommier from New York and architect Oliver Foster from Brisbane, Australia. They envisage a future in which fields of wheat, barley, or corn are grown in skyscrapers in the major cities of the world — on the roofs and all the floors below them too. These areas are also expected to have vegetable patches, orchards, deep-litter chickens, and water tanks for raising fish or shrimp.

LEDs will provide extra light, and the manure of the small farm animals will serve as fertilizer. Chemical pesticides will hardly be necessary, because the plants will grow in granulated material or in a solution with nutrients. This significantly reduces the consumption of water and fertilizer and makes it easier to keep ravenous pests away.

Instead of accepting the great expense, including high fuel costs, of trucking in food supplies from far away, buyers could get fresh vegetables, fruit, grains, and poultry from the high-rise farm around the corner. At the same time, these green oases would offer outstanding local recreation areas for stressed city-dwellers: the French-Belgian architecture firm of Vincent Callebaut has even designed floating green islands that could be positioned off the coast of the huge cities of Asia, or in New York's East River. It would be easy to supply these areas with water and energy, for example by means of wind power or wave energy plants.

A vision developed by architecture firm Vincent Callebaut. Floating farm islands and vertical gardens that supply fresh food to local residents.

Ten Square Kilometers of Farmland – Downtown

But aren't green skyscrapers much too expensive, because real estate prices in mega-cities are exorbitantly high? Not necessarily. Every city has enough unused land and vacant lots that could be filled with vertical farms. Despite their small area, these buildings can quite easily compete with large farms. That's because crops can be grown there all year round in green-houses, which means lettuce can be picked every six weeks, and even corn and wheat could be harvested three or four times a year. Despommier has calculated that, "in a 30-story building with a floor area of 60 hectares, the crop yield will be similar to that possible on a ten-square-kilometer farm."

Since about 6.5 billion people are expected to live in cities by 2050 – that's almost as many as the current population of the world – it is only logical to pursue concepts in which food is produced locally. In addition to achieving lower energy and transportation costs, another important benefit is the ability to show customers where and how the food they eat is produced. The general public is shocked by new food scandals virtually every few months, whether they pertain to spoiled meat, dioxin in salmon, hormones in pork, pesticides on vegetables, or antifreeze agents in wine. In more and more cities around the world consumers are therefore opting for the freshest food from known sources, and they are buying in stores that offer products from local farms.

Of course, many people will continue to fall back on cheap, industrial-ly-produced food to save money, but ultimately, consumers and politicians have the power to decide what is produced and with which methods. Yet as long as the average Western household's budget for tobacco and alcohol is twice the amount spent for fruit and vegetables, and more than the figure spent on meat or bread and baked goods, the priorities will remain misplaced – and that isn't because high-quality food is particularly expensive.

In 1970, an average employee in Germany had to work three times as long as today in order to buy a kilogram of pork chops, a liter of milk, or ten eggs. Food has become relatively cheap. However, most people today invest the money they have saved not in products from organic farmers but in bet-ter homes and in leisure activities. In the meantime, more money is spent on trips, entertainment, and hobbies than on nourishment, and a typical German household invests more euros per month in health care than in fruit, vegetables, and bread combined. If two pounds of coffee are cheaper than ten minutes in a solarium, no one need be surprised that the 25 million people who live from coffee cultivation in developing countries sometimes have to work in degrading conditions; that they are kept dependent on com-panies that dominate the market; that their wages are only a small percent-age of the price of coffee in the stores; that forests are being cut down for coffee plantations; and that too many harmful pesticides are sprayed on the crops.

But there are also some encouraging developments. For instance, the fair trade movement has become increasingly popular in recent years. In the fair trade system, various organizations issue quality labels for products like coffee, bananas, sugar, wine, and cotton goods. Although these products are usually priced higher than the world market price in each instance, the extra margin ensures that small local farmers and craftsmen earn an appropriate income and are not disadvantaged by large companies. In addition, a "Fair Trade" or "TransFair" label indicates that environmentally-sound methods were used during production and that working conditions, such as child labor and production conditions that pose health risks are not permitted.

By purchasing such products, any consumer can do his or her part for a globalized economy with a human face; approximately 1.5 million workers and farmers in Africa, Asia, and Latin America already benefit from fair trade. Although one can hardly assume that the majority of goods will be produced under fair conditions by 2050, the percentage that are will ultimately be determined by each individual consumer. And media organizations can do their part as well by persistently denouncing injustices and pressing for change. The example of keeping hens in laying batteries shows that change is possible. Though it used to be common, this practice of battery farming will be banned for good in the EU beginning in 2012 because of pressure from animal welfare activists and consumers.

Tracking Cutlets and Medicines

In the future, technology will help to follow foods "from the barn to the plate." It will let us know where an animal was born, where it was raised, and where it was slaughtered, so that a pork chop, turkey breast, or veal cutlet can be eaten without any fear that the meat is spoiled. To provide end-to-end monitoring of this kind, scientists will rely on electronic labels called "RFID tags." These labels have minuscule memory chips that contain important product information, and they can be read without contact via radio waves. In contrast to bar codes, RFID tags can be read from several meters away, even when the tag gets dirty.

Manufacturers of expensive electronic products and pharmaceutical producers also want to introduce tamper-proof RFID tags. There are several reasons for this. The pharmaceuticals industry, for example, currently assumes that one out of ten medicines is a counterfeit, and that pharmaceuticals worth 30 billion euros disappear every year. Electronic labels can store certificates of authenticity, and when they are combined with temperature sensors, they can also record temperature variation in order to prove later on that sensitive medicines were refrigerated continuously while being transported. All that remains to be done before these RFID tags can be deployed in large numbers is to lower their cost. Researchers are therefore currently developing tags that can be printed on films by the thousands, at which point they will cost less than one cent apiece. Scientist call this "polymer electronics," which means that electronic circuits are based on plastics.

Monitoring the global movement of goods will be absolutely necessary in the future, because such traffic continues to grow at enormous rates. Experts estimate that by 2025, it will increase by two thirds. One simple number illustrates how closely interconnected the world has become. In the last 60 years, the volume of global trade has increased 30-fold to 12.5 trillion euros per year at present. Of that total, the exchange of goods accounts for 10 trillion euros, and services account for 2.5 trillion. The volume of international trade is thus larger than the entire economic output of the European Union. Each year, Germany alone exports approximately one trillion euros worth of products, primarily automobiles, machines and chemical products. One in four jobs in Germany depends on exports. In addition, job conditions in Germany improve as German industry becomes more competitive in the globalized world.

From Swampland to a Forest of Skyscrapers

In Africa, most countries have so far not profited from the globalization of the economy, unlike the situation in East Asia and Latin America. China, for example, has increased its economic output five-fold since 1990 and thereby lifted 500 million people out of extreme poverty. At the same time, the number of first-year students at some 2,300 Chinese institutions of higher learning has also increased five-fold, to 25 million, and in the last ten years, the number of Chinese patent applications has increased six-fold. The "Middle Kingdom" is well on its way to becoming the most powerful economy of the 21st century. With regard to its economic output, China is now about one third as strong as the U.S., at roughly the same level as Japan, and just ahead of Germany. But analysts at Goldman Sachs are predicting that the Chinese economy will overtake the U.S by 2050 and will then probably be ten times as large as that of Germany. In the mid-1980s, there were hardly any buildings with more than 18 floors in Shanghai; today there are more than

6,000. Back then, the Pudong district was still a sparsely-populated swampland; today it is said to have more skyscrapers per square kilometer than any place in the world.

But of course, not everything that glitters in the sparkling cities of China is gold. The poor rural population in the west has, for the most part, not been lifted by the country's rising economic tides. That is one of the rea-

Chinese century. Some 25 years ago, this district of Shanghai was still a swampland; today it is home to the world's highest concentration of skyscrapers.

sons why 150 to 200 million migrant workers prefer to toil away on big construction sites of the mega-cities than till their fields at home. In the cities, there is a growing group of people who earn at least 10,000 euros per year and who now not only have their own apartments but, increasingly often, a car as well. According to the Chinese Academy of Social Sciences, the middle class thus defined includes 150 million citizens. In just 15 years, that figure could grow to 500 million.

And this is precisely where China has been concentrating all its efforts – on its own development. The new superpower is focusing not on expansion but on increasing the prosperity of its citizens. "We're no longer living in the age of territorial imperialism," says Kenneth Lieberthal, a China researcher at the Brookings Institution in Washington. "If China wants farmland in Africa, it just buys it." Because of its huge exports and highly-managed exchange rate, China now has foreign exchange reserves of between two and three trillion U.S. dollars. Of those reserves, the Chinese government has already authorized expenditures of many billions in order to buy foreign firms and acquire agricultural lands in Africa, South America, Australia, and the former

Soviet republics. It has also made large investments in Africa aimed at safeguarding supplies of raw materials such as oil and metals. In China, these economic interests take precedence over everything else, including the fight against terrorism and climate change, for instance.

The Global Finance Casino

It remains to be seen whether the policies pursued by China are really sustainable, or whether growing environmental problems and the demands of an educated middle class for more influence and democracy will cause structural upheavals in the country in coming decades. Nevertheless, China's example illustrates that if countries succeed in creating the right conditions for an economic upturn — from education and energy supplies to healthcare and transportation infrastructure — and if their citizens produce goods and services that are in demand elsewhere in the world, then globalization may have more advantages for them than disadvantages.

But this process of globalization cannot be allowed to take its course without regulation, as the recent economic crisis has shown. After all, unlike real products, money can be transferred to other countries in seconds, and a bank manager will always invest it in the place that promises the highest returns. In the U.S., for example, there was a long period in which even people who had no regular income were able to obtain loans for building homes and buying cars. These loans were then packaged together with those of more reliable debtors, divided up into groups, and repeatedly reshuffled until the risks were no longer easy to understand. Then they were sold to interested parties around the world with the promise of high interest rates: the risk was thus spread across the entire globe.

These transactions no longer involved real objects of value but were instead more akin to gambling in a huge casino. A new international financial system would be advisable. But it hasn't yet been possible to establish one. On the contrary, the billions of euros in assistance that banks obtained from governments at extremely low interest rates during the financial crisis were used by some not to extend loans to companies in dire need of them. Instead, they used the money to speculate on the stock and commodities exchanges, which made the crisis even worse, because now the prices for oil, gold, copper, and even foodstuffs rose too. Even worse, in order to cope with the crisis, many countries organized economic stimulus packages worth hundreds of billions of euros. The debts that must be paid off in the coming decades have thereby been driven to dizzying levels. These debts are often financed through government bonds for which relatively high interest rates are paid, depending on how creditworthy the government is in each particular case. Some countries, such as Greece, have in this way come to the verge of national bankruptcy. In order to save them, large loans and guarantees were once again needed, from which the banks and foreign exchange speculators stand to gain once more.

It is obvious that this sort of conduct in financial affairs is anything but sustainable. If politicians fail to overcome resistance to change in the financial community and fail to bring global monetary flows under better control, a number of additional crashes and economic crises will have to be overcome before 2050. And yet the solutions are plain to see: Lawmakers could introduce a tax on international financial transactions, demand that banks hold more equity capital, prohibit some types of transactions entirely, and introduce fees for high-risk transactions. The resulting income would flow into an insurance fund that would be available in the event of a bankruptcy.

Some measures have already been taken, but the laws that have been enacted so far to ensure more transparency and better regulation will hardly be sufficient. In 2009, German Chancellor Angela Merkel went a step further when she proposed the formation of a "Global Economic Council" analogous to the Security Council of the United Nations. And although this initiative has likewise come to naught, it may well be "retabled" during the next economic crisis.

A Very Different Kind of Bank

The Nobel Peace Prize laureate Muhammad Yunus demonstrates that there is another way to do things. Yunus, now 70 years old, had the pioneering idea of "rounding out capitalism by introducing social enterprises." At a conference in Berlin commemorating the 20th anniversary of the fall of the Berlin Wall, the founder of the Grameen Bank, a native Bangladeshi who obtained a doctorate in economics in the U.S., recalled how the walls fell in his own mind in 1976: "I looked at how normal banks behave and decided to do the opposite. Instead of distributing profits, we invest them in the company again. I take my bank not to the city but to the countryside. Loans go not to the men but to the women. I don't have branch offices that the customers have to visit. My employees go instead to the people in order to provide advice and collect interest on loans. And I don't need any lawyers, because I don't ask borrowers for any guarantees that they can repay the money."

In this way, Yunus has sparked a revolution in the system, and for this he is admired in much the way that Mother Teresa was when she founded the Missionaries of Charity and lived among the poorest of the poor in the slums of Calcutta. Yunus, on the other hand, does not think much of charity; he manages a commercial enterprise, but one that also does an enormous amount for people by liberating millions from poverty. "Gifts don't encourage anyone. They make people lazy," he says. At Grameen Bank, no one is given any gifts — unless you're prepared to describe trust and the chance to go from alms receiver to business partner as gifts. "That changes the psychology of people," says Yunus. "They can finally become active themselves and start managing their lives on their own."

Yunus invented the concept of microloans. His bank typically lends less than 100 dollars to people who have nothing but their capacity to work, and they can use the money to set up a small business. These businesses might involve sewing bed linen, mending shoes, making kitchen pottery, or running a chicken farm. The interest rates on the loans are relatively high

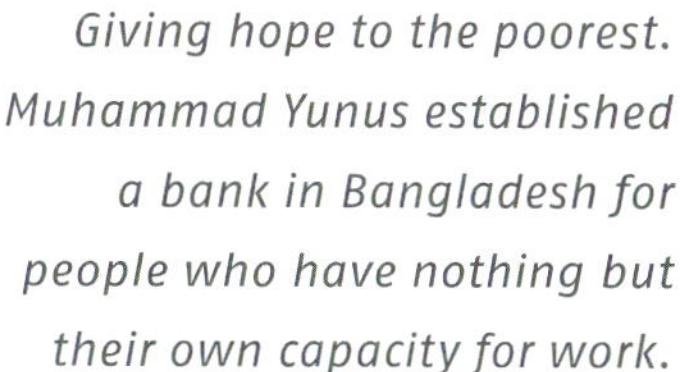

Giving hope to the poorest. Muhammad Yunus established a bank in Bangladesh for people who have nothing but their own capacity for work.

by western standards, but still far below the extortionate rates charged by private money dealers. Since borrowers are mostly women who are organized in village cooperatives, Yunus achieves repayment rates of 97 percent — for a simple reason. The social fabric is so strong in many developing countries that each woman wants to repay the loan because she would otherwise lose face in her village. Today, Yunus's employees serve about eight million families in 150,000 village centers. "Every month, we make loans for a total of about 100 million dollars," he says. And what's more, the bank even provides educational scholarships and micro-insurance against death or loss of wages.

When One Euro Can Buy a Pair of Shoes

Microfinance has had a broad impact. The United Nations and other institutions see it as an important instrument for fighting poverty. Worldwide, there are now probably over 70,000 microfinance institutes that follow the idea pioneered by Yunus. They have 100 to 200 million customers, and the number is growing all the time. In 2009, in the middle of the financial crisis, Yunus even exported his system into the center of capitalism. The Grameen Bank began making loans to customers in New York. In the first year, 600 women there obtained about 1.5 million dollars to build modest lives for themselves.

In Bangladesh, Yunus is now working with partners who, he says, "want to make a positive change in the world." Working with Danone, for example, he began marketing a yogurt product that tastes good and contains enough iron and vitamins for healthy child nutrition. "No one here asks about financial gain; we only ask how many children will ultimately profit from this," he says. He also persuaded Adidas to produce shoes for prices under one euro, because foot ailments are a large problem in Bangladesh. "But prices like that are only possible because we don't need any elaborate packaging or advertising," Yunus explains.

Taking stock of his experience, the charismatic revolutionary of finance makes a simple and encouraging observation: "Whether I talk with private persons or companies, I always find people who would like to do good things. It's not as though every kind of economic activity has to be an ego trip. I believe that social business still has a bright future."

The world of 2050 will need both of the solutions indicated above. In other words, it will need more transparency in global financial flows along with more control over those flows, in order to stop the gamblers who harm the whole world. And it will need more people like Muhammad Yunus who dare to pursue new approaches to "social business" and inspire others to participate with their passion and powers of persuasion. Only then will there

be any chance of giving the poorest people in developing countries hope for
a better life – one that includes fresh prospects not just in the big cities but
in the countryside too.

New York, January 2050. Thousands of video cameras, many of them mounted on the walls of skyscrapers, have to be checked. For the first time ever, the city administration is using newly developed mini-robots to do the job. Like oversized geckos, the robots climb up glass façades with their suction-cup feet. High above street level, the wind whistles through the hairs that cover their tiny metal bodies. These hairs were not added for aesthetic effect; they are piezo elements that transform the force of the wind into energy, thus recharging the robots' batteries. Thanks to their navigation systems and constant radio contact with each other and headquarters, the robots find every surveillance camera. They check the cameras for cracks and signs of wear or manipulation and replace old microchips with the latest versions. The cameras are supposed to "see" precisely, even from great heights, and alert human security guards to unusual occurrences. In cameras on lower floors, the robots also install, for the first time, software that enables them to compare the faces of passersby with images stored in a database. If there is a high degree of similarity with the face of a wanted criminal, this person is marked on the screens at security headquarters and the cameras then automatically follow the person's path. Big Brother is watching you...

LIVING SAFELY IN THE GLOBAL VILLAGE

The world has shrunk. Many local occurrences on Earth have an impact on the planet as a whole. The carbon dioxide blown into the air by a power plant in Russia recognizes no boundaries, nor does an uprising in Gaza affect only tiny Palestine. A virus that leaps from an animal to a human being for the first time in a backyard in China can cause an epidemic in New York a few weeks later — just as a computer virus smuggled into a data network by a hacker in India can infect a traffic management center in Brisbane or Minneapolis within minutes. How can people live safely in the year 2050 in spite of such challenges? How can they be protected from violence in densely populated cities and from attacks on their life-sustaining energy and data systems?

Current developments in energy networks are already moving in the right direction. That's because in the future there will be thousands more energy producers than there are today. Whether it's solar collectors on roofs, mini power plants in cellars, electric cars used as power storage units, biomass plants or wind turbines, there will be a great variety of energy generators that are closely networked with one another and make it possible to channel power in every direction. Ultimately, a kind of "energy Internet" will form, making the energy system far more secure than the one we have today, where a few major power plants could become the targets of terrorist attacks. In 2050 there will still be major power plants — for example, gigantic solar installations in North Africa — but the stability of the entire energy system will be less dependent on such installations. It will be similar to the Internet, where communication still functions even if some circuits are blocked, because the data is then simply channeled through other cables. In the power grids of the future, electricity will flow through alternative routes if individual installations break down.

In electronic data networks, hacker attacks, Internet spying, and cyberattacks by terrorists are a new form of criminal activity. For example, in the U.S., hackers have stolen the plans for a fighter plane and spied on hundreds of high-tech companies. Inspections of power grids have also revealed external programs that would have massively disrupted operations if they had been activated. Even Google became a victim when Chinese hackers spied on the e-mail accounts of political activists.

And in the summer of 2010 news sources reported on an attack that software experts called an "entry into the age of cyberweapons." For the first time ever, somebody had succeeded in producing malicious software, known as a Trojan horse, that is so precisely targeted that it can do damage only in very specifically-defined industrial installations. This Trojan horse, called Stuxnet, disseminated itself via USB sticks, infected thousands of computers that were using Windows software, and searched there for certain industrial control systems, within which it would then change setting parameters such as pressure, flow rates, temperature or the rotation speed of centrifuges. Many experts believe that the objective was to sabotage a Iranian uranium enrichment plant, but it may never be known whether or not this aim was achieved.

Another unanswered question is who designed this software. These were certainly not ordinary criminals, because the effort required was enormous and obviously only one very specific installation was targeted, while all other infected systems remained unharmed. A team numbering between five and ten specialists probably spent well over half a year developing this Trojan horse, and they very skillfully exploited errors in the Windows system and stole digital signatures — activities that may have cost millions. In fact, experts believe that a "digital first strike" of this kind could only be carried out by intelligence agencies — such as those of the U.S. or Israel.

Since that time, security experts have been extremely busy. Removing "malware" such as Stuxnet from a computer is basically not a problem

– the real difficulty lies in finding such unique weapons in the first place and protecting oneself from them in the future. Today firewalls are used to keep unauthorized users out of protected subnetworks, and experts forecast the future development of "immune systems" for computers. These would be smart programs that learn from experience; in case of an attack, they would fall back on the knowledge stored in many databases in order to repel intruders. In addition, anti-virus specialists are trying to develop electronic signatures that cannot be counterfeited – for example, images equipped with digital watermarks that will protect them from manipulation.

Preventing Undetected Eavesdropping – with Quantum Physics

Those who want to save or send data securely use encryption. However, even good codes can be cracked with high-powered computers. But now scientists have developed a process that enables absolutely tap-proof communication. It is based on an effect that Albert Einstein dismissed as "spooky action at a distance," even though he himself had discovered it – the so-called entanglement of quantum systems.

When specially prepared twin pairs of photons are created, one photon always "knows" the state of the other without any delay in time and across any distance, even that of the entire universe. What's more, every disturbance of one of the photons is immediately registered by the other. This means that it would be physically impossible to eavesdrop unnoticed on a communication line that works with entangled quanta.

Researchers working with Anton Zeilinger, a professor at the University of Vienna, Austria, have succeeded in setting up this kind of tap-proof data communication. Using 1.5 kilometers of glass fibers, they transferred 3,000 from Bank Austria to the Vienna City Hall, thus demonstrating one of the first possible applications of the system. Not only banks and public authorities but also police departments and the military are interested in secure communication lines of this kind. Private individuals could also benefit from such systems by 2050, when glass fiber connections reach into every household and we can simply stick a quantum cryptography device into our computers just as we do with USB sticks today.

Cameras that Detect Abandoned Suitcases

Security will be a dominant issue in the information society of the future — and not only on the Internet. Analysts expect that the global market for security solutions will nearly double within the next ten years. In cities, the primary example is the installation of surveillance cameras in subways, commuter railroads, and public areas. London alone has installed approxi-

Big Brother is watching you. There are about a million surveillance cameras in London. The best of these systems can already independently recognize abandoned suitcases and other potential dangers.

mately a million such cameras. They are everywhere: on buildings, in streetlights and traffic signs. They can't prevent crimes, but at least they make it possible to solve them faster. For example, after the terrorist attacks in London in July 2005 the analysis of videos recorded by the cameras quickly put the authorities on the terrorists' trail.

Processes of this kind would not work without computer support, because security personnel cannot analyze millions of images themselves. A U.S. study has shown that even when watching only two surveillance monitors, a human viewer will miss up to 45 percent of all the activity in a scene after 12 minutes and 95 percent of what's going on after 22 minutes. By contrast, cameras and computers don't get tired. In laboratories, experts are designing smart image processing programs that can analyze scenes themselves and notify security personnel only if unusual events occur. The best programs can recognize abandoned suitcases in airport terminals, cars that are driving through tunnels in the wrong direction, or people who are standing dangerously close to the rails in subway stations. At night they can distinguish between a dog and a human being in a garden on the basis of the size, shape, and speed of the intruder. After a camera has registered a suspicious person, automatic motion analysis and object tracking can be initiated. This means that if the target moves into another camera's range of vision, the software automatically passes along the information so that the path taken by the intruder can be precisely tracked.

The one element that will not function reliably enough for many years yet is the use of video images to automatically search for criminals. There are simply too many possible variations in terms of the angle of vision, lighting, level of occlusion and changes to a face by means of a beard, glasses or a different hair color. But thanks to increasingly powerful computers and software, by 2050 computers will at least offer very good probability rates that a given individual corresponds to a wanted person. That would increase security, but it could also become a nightmare if it results in security authorities persecuting innocent people and investigating every aspect of their lives. In the future, those who correlate the data gathered by surveillance cameras with an individual's telephone, Internet, energy use, credit card, and bank data will receive an almost complete profile of that person. In principle, technical progress will open the gates to a perfect police state.

Looking for Faces with a Cell Phone Camera

The door to the nightmare of constant surveillance is opening as the public's resistance is eroding. Various surveys show that most Londoners would have no objection to being tracked several times a day if this is done to promote public security. The situation in Germany is similar. Whereas in the early 1980s protests against a planned census initially prevented it from being carried out, many people today are voluntarily making lots of personal data public, for example in customer discount campaigns, contests on the Internet, and detailed blogs on Facebook and other social websites. "The Internet forgets nothing" — as some people realize only after a potential employer has checked the Web and discovered their embarrassing photos of a beach party in Croatia or a very private video made in Mallorca.

In a few years image processing software may reach the point where anyone can expect to be quickly checked out in a café or a disco. A click on a camera-equipped cell phone could activate a program that searches all the databases in the Web and delivers the most probable matches. Seconds later, the cell phone owner would be able to study a target person's profile, including his or her preferences, hobbies, address, curriculum vitae, friends and possibly even what he or she did in the past few days or has permanently saved on the Internet. "If people voluntarily give up their privacy this way, one day it could really be at risk," says data protection expert Marc Rotenberg from the Electronic Privacy Information Center in Washington, D.C. "The protection of our privacy, as of all other social and political values, basically depends on the will of the public."

Security and privacy need not be at odds as long as political control mechanisms are work properly and are up to date, according to Rotenberg — in other words, as long as the authorities make sure that stored information is not passed on to third parties in an uncontrolled manner and that people have the right to access and correct data that affect them. In addition, as little data as possible should allow users to deduce an individual's identity. For example, in the future, access to buildings or events will increasingly be authorized via fingerprint, voice or face recognition technology. However, these data should merely generate a code that indicates that the person in question is authorized to enter the building — without divulging his or her identity. This degree of anonymity is the same as that of the purchaser of a movie ticket. The buyer holds a document, such as a driver's license showing that he or she is over 18 and thus authorized to see the film, but the ticket checker certainly doesn't know the identities of the individual moviegoers.

Identification in 3-D. Distortions of the colored bands enables a computer to identify the three-dimensional shape of a face in fractions of a second — in a very secure process.

A particularly secure new process is that of 3-D face recognition. Here, a grid consisting of infrared bands is briefly projected over the individual's face. A computer uses the distortion of these bands across the cheekbones, nose, and forehead to calculate shapes in fractions of a second — with an accuracy measured in tenths of a millimeter. Unlike a simple two-dimensional image, which basically could be replaced with the photograph of another individual, the three-dimensional shape of a face is very difficult to reproduce. The combination of such a 3-D facial image and a fingerprint is practically impossible to counterfeit — and in the future this will be ideal for applications such as safeguarding access to particularly sensitive offices or automatic teller machines.

Sensors that Identify Explosives by Smell

Concealed weapons or explosives can be detected by means of long-wave heat radiation, which is known as terahertz waves — even if they are being carried under clothing. That's why these scanners, which are currently being installed in many airports, are being called "naked scanners" by the media. In order to safeguard privacy, in Europe these scanners are to be used only to create an abstract representation of the human body. Explosives can also be detected by means of gas sensors. For example, The European Aeronautic Defence and Space Company (EADS) has developed an ion spectrometer that "inhales" the odor molecules coming out of a suitcase and filters out the molecules of any explosives contained in it. A single molecule of an explosive within a sample of a thousand billion molecules is sufficient to trigger an alarm. This degree of sensitivity is 30 times as fine as that of a dog's nose. However, the noses of trained sniffer dogs can detect a much broader range of substances, including drugs, for example.

Attacks on drinking water supplies would be particularly dangerous. In 2005, for example, an unknown person submerged three canisters of herbicide in Lake Constance, on the northern foot of the Alps, very close to a pumping station for drinking water. Thanks to an anonymous letter from the person responsible, divers found the herbicide at a depth of 70 meters and were able to prevent it from entering water distribution networks. But the incident highlights how great the risk is for cities and regions. In order to be able to sound an alarm early, scientists are developing sensor systems that measure the activity of certain biological enzymes. A decrease in their activity would indicate that insecticides or chemical weapons are present.

In the future, other systems will be able to extract dangerous bacteria from the water. And there are already solutions to deal with even one of the most dangerous scenarios: a bomb containing radioactive material. The port of Rotterdam, which handles 400 million tons of goods annually and is therefore one of the world's largest harbors, is playing a pioneering role in these efforts. There, every container must be transported via truck through one of 35 gates that detect radioactive isotopes.

In the world of tomorrow, many of the systems that are still separate today will be more closely networked. For instance, security services will be linked with transportation, energy, communication, and healthcare systems. In one example, if a fire breaks out next to a subway station, the cameras and smoke sensors installed there could send an alert not only to the fire department, but also to the subway control center, which could then reroute trains as necessary. On the streets, an automatic alteration of traffic light phasing would direct cars away from the danger zone and simultaneously keep roads clear for fire trucks, police cars, and ambulances. First responders would receive all relevant information on mobile devices, including images from cameras at the site of the fire and information on how to navigate safely in spite of smoke. Gas conduits would also be automatically shut off. Last but not least, hospitals would be warned so that they could prepare in time for the arrival of the injured.

Psychology and Technology

Nonetheless, even in the future there will be no comprehensive security measures against criminal activity and terrorists. For example, in the case of the failed attack on a U.S. airplane in December 2009 even terahertz scanners might not have detected the explosive in question, because the terrorist from Nigeria smuggled it into the plane in the form of a fine powder distributed throughout his underwear. But even more problematic was the fact that the security services had not managed to put together the father's warnings about his increasingly radicalized son, entries in U.S. intelligence reports, and the terrorist's own flight plans into an alarming overall picture. The situation is very different at the Israeli airline El Al, which is regarded as the safest in the world even though it is the one most at risk. Its security system relies not only on technology, such as built-in missile defense systems, but also — and particularly — on specially trained personnel. All passengers are interviewed using a special questionnaire, and close attention is paid to their behavior in order to identify possible assailants. This approach is justified by its success. Over the past 40 years there have been no terrorist attacks aboard El Al airplanes.

In order to stem international terrorism, more than pure military technology is needed. The ultimate goal is to deprive terrorists of backing from those members of the general public who support them financially and morally. In one case, this approach has already been successful. The terrorist group known as the Red Army Faction in Germany was defeated not so much because of the imprisonment of many of its leaders, but because it lost the sympathy it had originally received from certain groups within the population. As a result, the last RAF members announced in 1998 that they were dissolving their organization. Receiving a letter to this effect from Osama bin Laden's al-Qaida would be a dream come true for security experts and the public.

But a lot would have to happen in order to dry up the continuous flow of support for terrorists. These measures range from the creation of functioning governmental structures in Iraq and Afghanistan to ending the civil wars in Somalia and Yemen, bringing about peace talks between the Palestinians and Israelis, and protecting the minorities in Indonesia. The basic goal is always to eliminate oppression, poverty, and discrimination and to promote education — to teach people humane values rather than the hatred that is preached in some environments even by renowned authorities. It will take a long time to achieve success in this area, but this is one of the vital political and social tasks of the 21st century. That's because the fundamental evil that leads to suicide attacks lies not in religions but in the social conditions in which people are forced to live. If these conditions were improved so as to make life more worth living, probably very few people would consider death a desirable goal.

Affordable Water for All

That's why the Millennium Development Goals of the United Nations, such as overcoming poverty and starvation, are essential milestones on the long road toward world peace. Take water, for example: According to the UN, approximately 80 percent of all infectious diseases can be traced back to dirty water that has been polluted by pathogens. Every year 1.8 million people die of diseases connected with diarrhea, and 90 percent of them are children under five years of age. In India alone, 1,000 children die every day.

Providing a supply of clean water is a Herculean task, as 60 percent of the world's usable reserves of drinking water are located in only ten countries. About a billion people have no access to clean drinking water, and some 2.4 billion have no adequate wastewater disposal system. Even in China, which is on an economic upswing, hundreds of millions of people must cope with an inadequate supply of drinking water. More than half of China's rivers are so polluted that they must be ruled out for this purpose. In addition, of the 20 places with the poorest air quality in the world, 16 are in China. The Chinese government estimates that environmental pollution causes about 400,000 deaths per year in China. "We pride ourselves on being the workbench of the world, but if we're not careful we'll end up being the planet's biggest garbage dump," was the warning expressed in interviews by Pan Yue, Deputy Minister of China's Ministry of Environmental Protection.

10,000 liters of drinking water per day. SkyHydrant transforms muddy slop into high-quality drinking water, and nano pores even keep bacteria out.

But that doesn't have to happen. For the drinking water problem at least, solutions are in sight. "From a technical viewpoint, we are now able to provide everyone on Earth with clean water, even in the most underdeveloped regions of the Third World," says Rhett Butler, the originator of the Skyjuice Foundation in Australia. Butler is the developer of SkyHydrant, a device as tall as a man that can convert 10,000 liters of dirty water per day into absolutely pure drinking water. What's more, its design is so simple that it can be set up within ten minutes and operated by just about anyone without any lengthy instructions. The system requires no electrical energy or processing chemicals, will last for ten years with proper maintenance, and at an annual cost of 20 to 30 euro cents per person for an annual supply of drinking water, it is affordable even for the poorest communities.

Its design is based on high tech, but the way it operates is simple. Water is forced through 20,000 extremely fine fibers that look like meter-long spaghetti but upon closer inspection turn out to be hair-thin tubes with tiny holes in their walls. These long straws are membranes that permit water molecules to pass through, but no particles that are bigger than 100 nanometers – in other words, one ten-thousandth of a millimeter. No suspended solids or bacteria can get through, and even viruses, which are much smaller, are trapped because they are always bonded to other organisms. The natural pressure of water pouring downward from a height of two meters is sufficient to force the water through the membranes. The fibers are cleaned in a very simple mechanical manner when they rub together as the user shakes the whole bundle of fibers by means of a lever.

The SkyHydrant is currently being used in India, China, and along the Gona Reservoir in Kenya, whose water used to be drunk by the inhabitants of the surrounding villages – but who consequently suffered from diarrhea, cholera, and typhoid fever as a matter of course. Today a small windmill drives a pump that forces water from the reservoir into the SkyHydrant. There it is converted into drinking water that exceeds even the high quality requirements of the World Health Organization. "We're talking with microloan banks that could sell the system to water kiosks in the future," says Butler. Such water kiosks are operated in developing countries by small-scale entrepreneurs to provide people with clean drinking water.

Wastewater – a Cult Drink in Singapore

Membrane technology similar to the kind used in the SkyHydrant is also utilized in the large purification plants that provide cities like Beijing and Singapore with clean water. For example, Singapore, which has a population of five million, has set up a membrane-based purification plant equipped with an ultraviolet disinfection system that recycles the city's wastewater into more than 200 million liters of clean water per day – a volume that covers one fifth of the city's water needs.

The purified water, which is known as NEWater, is mainly used by the semiconductor industry and other organizations that require an exceptionally pure product, but it has also become a cult drink in Singapore. It is even consumed by Singapore's prime minister, who serves it to state guests.

Drinking recycled wastewater is certainly not everyone's idea of a treat, but for the inhabitants of future megacities it is probably inevitable in order to make optimal use of scarce resources. In any case, Singapore has definite plans to become the worldwide center of expertise for water technologies — for example, by issuing a call for tenders for a million-dollar research project to develop the best technology for halving the cost of purifying seawater to produce drinking water. The contract was won by a Siemens concept that, instead of desalinating seawater through energy-intensive heating and vaporization processes, involves channeling the water through an electric field that removes salt ions. This reduces energy consumption per cubic meter of water from the ten kilowatt-hours (kWh) common at conventional facilities by 85 percent to just 1.5 kilowatt-hours. Even the best methods used to date still consume twice as much energy. The new concept will now be refined to the point of market readiness so that it can be widely used in coming decades.

New Rice for Africa

In addition to providing a supply of drinking water, eliminating starvation is also vital to ensuring a peaceful world in the year 2050. A billion people throughout the world do not have enough to eat; 640 million of them live in Asia and 300 million in Africa. More than half of these people are farmers — and that's particularly deplorable, because it's obvious that even the people who produce food can not always make a living at it. In the future we must succeed above all in boosting food productivity, for example through irrigation, better seed and fertilizers, and careful measures that enhance soil fertility, such as alternating the cultivation of different kinds of plants. At the same time, we must reduce the enormous amount of waste resulting from poor transportation and inadequate warehousing. Studies show that in many cases between 40 and 60 percent of harvests is destroyed in this way.

However, starvation does not have to be regarded as an inevitable fate. For example, microloans have helped many farmers in Bangladesh to put their lives on a solid economic foundation (see p. 142). Malawi supports its farmers very successfully with a system of vouchers for subsidized seed and fertilizer. As result, the country's agricultural production has doubled. The development of New Rice for Africa, also known as NERICA, by scientist Monty Jones from Sierra Leone has also turned out to be especially helpful in Africa. NERICA is a hybrid that was created by crossing an African rice variety that is very resistant to drought and local pests with a very high-yield Asian variety. NERICA now enables farmers to harvest 2.5 metric tons of rice per hectare — and if they use fertilizer they can raise that figure to five tons. By contrast, the purely African rice variety yields only one metric ton per hectare.

Rice has become a staple food for 240 million people in West Africa. However, to date most of it has had to be imported. As a result, in 2008, when the rice harvest was rather poor and the Southeast Asian countries began to hoard their stockpiles of rice, speculators forced prices to triple in some cases and people in Africa consequently starved. This led to violent unrest in Cameroon and Senegal, and even in Haiti and Indonesia. It's not hard to imagine what this could lead to when the poorer countries of the world need to feed over two billion more people by 2050. It will be a difficult task to find the right balance between locally-grown grain varieties such as NERICA and inexpensive imports from countries where some foods can be produced more cheaply and even more plentifully. And intermediaries must also be strictly monitored so that speculation with foodstuffs does not put thousands of human lives at risk.

Some countries are offering assistance. Russia, for instance, has calculated that it could cultivate enough additional agricultural land to feed 450 million people — three times the country's current population. But unfortunately, lots of agricultural land is simultaneously being lost in. For one thing, it is being used by many livestock growers — who, moreover, use grain as animal feed. For another, competition is growing between the dinner plate and the gasoline tank — in other words, when agricultural production is used not for food but to produce biofuels (see p. 56). "Agricultural land cannot be increased at will," said Stefan Marcinowski, a member of the Board of Executive Directors of the world's largest chemical company, BASF, in an interview. "In view of population growth, we have to double agricultural productivity in the next 20 to 30 years." He believes that genetic engineering is the major development that will help to boost productivity.

Producing Bacterial Pesticides with Genetically-Modified Corn

Genetically-modified plants, including corn, cotton, soybeans, and rapeseed, are being raised in 25 countries on almost ten percent of the planet's total agricultural land. In the U.S., four fifths of all corn varieties have already been genetically modified. One of the goals is more effective protection from insect pests. In this process, a genetically-modified plant produces a preform of a bacterial toxin that is harmless to human beings. This substance is transformed into a toxin in the intestines of butterfly caterpillars such as the corn borer and causes this insect pest to die of starvation. Pesticides containing such toxins have been sprayed on crops for decades, and their use is even permitted in organic farming. The advantage of inserting these substances directly into a plant by means of genetic modification is that this approach, unlike spraying, affects only the caterpillars, because they are the only ones feeding on the corn. In other words, other animal species are preserved.

The cultivation of another variety of corn that is equipped with a different bacteria gene will be authorized starting in 2012. The gene is expected to help the plant cope with stress situations such as drought. This corn variety could therefore also be planted in areas where corn cultivation is still impossible today. And in the laboratory researchers have made some progress in their pursuit of a key vision of the future. They intend to modify cereal plants in such a way that they can absorb nitrogen directly from the air, just as beans and peas do. This would drastically reduce the need for fertilizers.

Marcinowski has no doubt that in the future genetically modified plants will be raised even more widely than they are today, for example to produce oils with valuable ingredients or disease-resistant potatoes. This prediction may be correct in worldwide terms, but in Europe green genetic engineering will continue to have a difficult time because of widespread public skepticism in Germany, Austria and France. In the past 12 years, the cultivation of 150 genetically modified plant varieties has been approved throughout the world, but only two of these varieties have been permitted in the EU. Many Europeans are concerned that genetically modified plants could, for example, form hybrids with conventional plants, cause new types of allergies, or increase the resistance of pathogenic bacteria to antibiotics.

However, to date, studies of green genetic engineering have not discovered any proof of such health risks. The rejection of this technology is therefore due more to the fact that few citizens of wealthy nations see the advantages it would bring them, whereas its value is much more obvious in developing countries. In a survey conducted in Japan and France only 22 percent of respondents said that the advantages of genetically modified food plants were greater than the risks involved. In India and China that figure was considerably higher, at more than 65 percent, and it was highest in Cuba and Indonesia, at approximately 80 percent.

S.M.A.R.T. Products from Developing Countries

The supply of clean drinking water and healthy food is important, but it's still not enough to ensure that we will have a peaceful world in 2050. People also need a hopeful economic perspective — the opportunity to improve their living conditions. In the rural regions of sub-Saharan Africa, India, and Southeast Asia, approximately 1.6 billion people still live in what is practically an age of wood. In other words, they have no electrical power and use wood, straw, dung, and often garbage for cooking and heating. According to the World Health Organization, the pollutants produced by traditional hearths cause the premature death of 2.5 million people annually — far more than are killed by malaria.

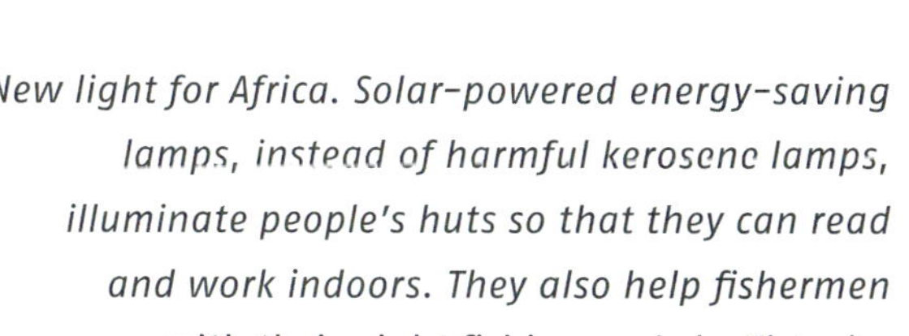

New light for Africa. Solar-powered energy-saving lamps, instead of harmful kerosene lamps, illuminate people's huts so that they can read and work indoors. They also help fishermen with their night fishing on Lake Victoria.

And yet it isn't at all difficult to generate environmentally-friendly energy in these regions, whether it's with small windmills, water wheels or solar cells. All the same, in the past, quite a few solar cell installations in the bush have expired due to lack of maintenance. Today, more emphasis is being placed on getting local people to participate in these measures. For example, micro loans are helping merchants in a Bangladesh village to purchase a solar power facility, complete with maintenance, and supply their neighbors with electricity for lamps, radios, and cell phones.

Through the Lighting Africa project, the World Bank aims to provide the huts of 250 million people in Africa with electric light by 2030. Lack of lighting is one reason why children in Africa and Asia can't study at night, which makes it more difficult for them to get a higher education later on. Thanks to the project, adults will also be able to do handicrafts indoors after sundown. Osram is one of the industrial partners of Lighting Africa. In the first project of this kind, it is substituting millions of energy-saving lamps for incandescent light bulbs in Africa and Asia. In exchange for financing the project, it will receive CO_2 emission rights that it can negotiate with in climate exchanges.

Energy-Saving Lamps from Solar Service Stations

In another project, Osram has installed solar-powered electrical service stations on Lake Victoria in Kenya. Here fishermen can rent energy-saving lamps, complete with rechargeable batteries, and have them charged with solar power. These lamps light the fishermen's homes and are also being used at night to attract the small sardine-like fish that live in the lake. The fishermen previously used kerosene lamps, which put them at risk of fire and gave off harmful vapors. The new type of illumination eliminates these risks, gives more light, and is even cheaper than kerosene. In addition, the solar cells are also used to operate a water purification system that uses microfilters and ultraviolet disinfection equipment. This is an excellent deal for people in Kenya. They receive environmentally-friendly lighting, healthy drinking water, and new opportunities to earn a living. Under the slogan "Lighting a Billion Lives," a similar program is running in India with the long-term goal of providing a billion people with solar-powered lamps.

Indian economist C. K. Prahalad believes that there is still tremendous potential "at the bottom end of the economic pyramid." Today approximately four billion people live on less than US $10 a day. And yet, according to the analyses of the McKinsey Global Institute, in India alone almost 300 million people could make the leap out of poverty by 2025. According to studies conducted by Goldman Sachs, about 800 million people in Brazil, Russia, India, and China — the BRIC states — have an annual income equivalent to at least US $6,000. Within the next decade, the number of people constituting this "middle class" of the BRIC states could double to include 1.6 billion people. These people don't need any high-tech systems. Instead, they need S.M.A.R.T. products (the term stands for "simple, maintenance-friendly, affordable, reliable, timely to market") that are tailor-made to meet their requirements.

There are many solutions of this kind that can be produced in developing countries and emerging economies. They range from small electric scooters to affordable cars, mini power plants that use coconut shells to produce electricity (see p. 56), inexpensive video cameras for monitoring industrial processes, and robust tomography systems that can even be sold as inexpensive auxiliary equipment to clinics and doctors' offices in the U.S. and Europe. And here's another example: In Indian villages, the heart frequencies of fetuses are being monitored not with complex ultrasound equipment but with simple portable acoustic microphones. For women who are having a problematic pregnancy, this is an important auxiliary method to reduce the risks for unborn children and for the women themselves. All in all, tremendous opportunities will be opening up for small-scale entrepreneurs and companies if they understand how to develop S.M.A.R.T. products of this kind. At the same time, they are a springboard that entire national economies can use to catapult themselves out of poverty. And for the human race as a whole, this is the most promising route to a more peaceful and secure world on the road to 2050.

Edinburgh, February 2050. It was an unusual job, even for Open Innovative, the market leader in global knowledge networks. The assignment, to be completed within 30 days, was to design a car seat that can separate itself from the vehicle it is installed in and continue to travel on its own, through a shopping center, for instance. The seat was to be controlled by voice and by joystick, and its environmental impact over its entire life cycle had to be zero. It took Open Innovative 48 hours to look through its pool of half a million specialists and form a core team consisting of environmental engineers and automation experts from France and Germany, production planners from Great Britain and China, software developers from Brazil and India, and 3-D simulation experts from the U.S. Standard components could often be used for image processing, radar sensors, and navigation. For other parts, new solutions had to be developed, mainly in order to meet requirement for biodegradable plastics and materials that can be recycled without loss of quality. This was possible only after locating suitable experts at the University of Copenhagen. After that, everything went like clockwork. All the details were simulated in 3-D, including the operation of the car seat, its robot-controlled manufacturing, and the recycling of the finished product. After 28 days, Open Innovative was able to deliver all the data to the customer — thus setting a new record.

WHO WILL DO TOMORROW'S WORK?

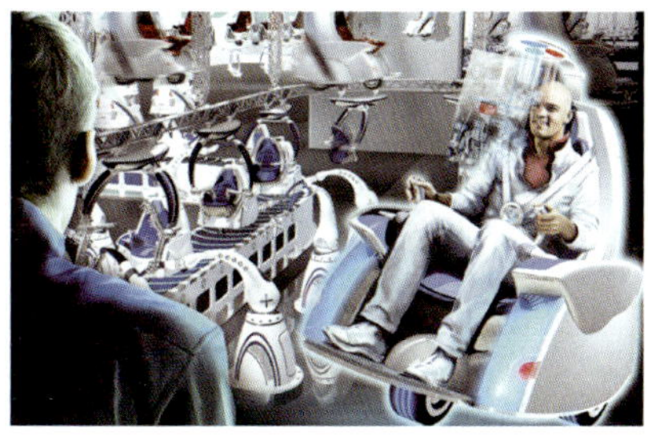

It's like a scene from a nightmare: enormous stamping pistons, interlocking gear wheels, hissing valves, dials with needles going crazy, flashing lights, and in the midst of it all are people like machine parts, anonymous in the crowd and occupied with jobs whose purpose is unclear. "Deep underground lay the city of workers." This is how Fritz Lang's film *Metropolis* introduces the hellish world of the slave-like workers who make it possible for a small upper class to live a life of luxury. This film, which had its premiere in 1927, has inspired countless masterpieces of science fiction and fantasy, from *Blade Runner* to *Star Wars*, *Superman*, *Terminator*, *The Fifth Element*, and *Matrix*.

It's not only the aesthetics of *Metropolis*, with its soulless city of the future, that have fascinated later filmmakers; it's also the timeless questions that it raises. How can work be organized so that it provides everyone with a life befitting a human being? Are machines our servants, or will they destroy us? How much like human beings can robots be? These questions have been asked for thousands of years, and they may still be asked a thousand years from now. In the Jewish Kabbalism of the 11th century, for example, there is mention of an artificial creature known as a golem, and in *Metropolis* the inventor Rotwang creates the machine-woman Maria. In 1921, the plot of a Czech play involved artificial beings known as "robota" who performed compulsory labor. And the same primeval fear continues to emerge today. People cannot control their industrial slaves; robots break free and annihilate humanity. In the film *Blade Runner* from 1982, the question of how much robots can resemble human beings is then carried to an extreme: Androids and people have become practically indistinguishable.

The question of which laws robots and all other artificially-created tools must follow so that human beings suffer no harm through their actions or inaction was investigated by the author Isaac Asimov. But although his laws of robotics sound simple and seem to be formulated without ambiguity, he often succeeds in constructing situations in which these rules can, or even must, be broken. One lesson is that a future coexistence of people and autonomous artificial creatures will not be easy, even if these robots do nothing more than follow clearly-defined rules — not to mention what would happen if they were to react intuitively or emotionally to situations, or even develop the first stages of consciousness.

But researchers have not yet reached the point where they can create emotional robots, and that may be a good thing. On page 115 you will find a description of what kinds of household jobs could be performed by a personal butler made of steel by the year 2050. And what robots with feelings could do is related on page 193 and the following pages. In industrial plants, on the other hand, engineers take a very pragmatic view. Robots help to manufacture products faster and more cheaply, and they are supposed to help companies respond to customer needs as flexibly as possible. The objective is customized production. In the auto industry, for example, the variety of special options available to customers means that any individual car on the production line is likely to be different from the next, and in tomorrow's world that will be true of many more products, from furniture to articles of clothing.

In the clothing sections of department stores, there will be scanner booths that use a laser beam to identify the precise shape of the customer's body in minutes; some trailblazers have already tested the first machines of this kind. The underlying idea is that once their 3-D data is scanned in, customers can "try on" whatever they like at home via the Internet. On the screen they would see a figure that matches their own down to the smallest detail. They would be able to position this avatar, turn it around, find the perfect clothing under different lighting conditions, and order it right away — tailored to their personal measurements and presumably manufactured by robots that react flexibly to new orders.

Testing a Factory before Breaking Ground

Computer intelligence is increasingly defining the phases of industrial production. For example, many products can already be simulated, tested, and improved on computers before a single screw or component is manufactured. The same applies to production processes. Entire factories can already be portrayed in three dimensions, and their processes can be optimized before they exist physically. This can greatly accelerate production, because many defects can be identified and eliminated in advance, and because quite a few variations can be tried out in the virtual world beforehand, which saves time and money. Nevertheless, there are still some processes that are difficult to simulate, such as combustion processes, with their multiplicity of complex physicochemical reactions.

Virtual before real. Trains are first designed and tested in 3-D computer simulations that include every detail before the real thing is built.

As a rule, simulations are always difficult when different types of processes must be combined, for example when the flow of electrical current in a component must be simulated together with associated heat flow, material fatigue, and vibrations. They are also difficult in the case of complex biochemical processes such as the effects of medicines. And the same is true when human behavior enters the picture, as in simulations of passenger flows at train stations or crowds at soccer stadiums.

Furthermore, simulations are difficult when completely disparate dimensions have to be dealt with. For example, today it is possible to simulate the behavior of several thousand atoms. Materials scientists can also analyze the properties of materials on a centimeter scale and smaller, down to the size of grains measured in thousandths of a millimeter. But bringing both scales together is something that still far exceeds the capabilities of the most powerful computers. Researchers would have to calculate the movements and the reciprocal influences of billions of atoms merely in order to reach a grain's dimensions measured in thousandths of a millimeter.

In 20 or 30 years, however, computers will be thousands of times more powerful than they are today and thus will be capable of solving many of these problems. And for all the other problems, one must bear in mind that an intelligent approach is often more important than computational capacity. For any particular problem, scientists must develop the right models — those that best approximate real behavior, even if they are unable to reproduce it in every detail.

A Society that Produces Zero Waste?

But how can it be possible to make products for nine billion people in 2050 without doing too much harm to the environment? This will obviously not be possible with the current economic system. In order to understand that, we need only look at the Pacific Ocean. Between California and Hawaii there is a floating carpet of garbage that is twice as large as Germany, consisting of 100 million tons of small pieces of plastic waste. No one likes to imagine what the oceans of the world might look like in 40 years. But there is an alternative. In principle, a life without garbage is possible. This is demonstrated by a species that metabolizes as much energy as 30 billion human beings but produces no garbage at all: ants. There are at least ten quadrillion ants on earth, but ants do not behave like individuals. They form colonies of thousands or even millions of individuals that take over jobs like nest building, taking care of the brood, finding food, and defense. They communicate via scents or by touching things with their antennae. They have been very successful with their strategy of social cooperation for many millions of years. Researchers call this "swarm intelligence."

The chemist Michael Braungart regards ants as a model. "They produce only nutrients and no waste; everything that they take from nature is returned as a product of metabolism," says Braungart, a former Greenpeace activist and co-founder of Germany's Green Party. "If we were to act as intelligently as ants, we could grow to a population of 30 billion, and other living things would still be glad we're around. Instead of being harmful organisms, we would turn into useful ones." Together with the architect William McDonough, Braungart hopes to teach people a little of this intelligence. They have developed a concept that turns the prevailing opinion of environmental researchers upside down. It is called "cradle to cradle," in contrast to the usual "cradle to grave," which describes the life cycle of products from their design to their manufacturing, use, and ultimate disposal. "We have to get the word 'waste' out of our heads," says Braungart. His motto is "Waste is food."

Working with McDonough, Braungart, who is now 52 and a professor of process engineering, coined the term "eco-effectiveness." Products that are eco-effective, he explains, are those that are either returned to nature as biological nutrients after their use, or those that serve as "technical nutrients" – in other words, they can be reused for manufacturing technical products. So far, however, environmental researchers and companies have mostly tended to pursue eco-efficiency rather than eco-effectiveness, which means that they hope to consume the smallest possible amount of resources and produce the least amount of pollutants. This is completely different from the world of nature, where an intelligent sort of wastefulness prevails. "For million of years, plants have been producing utterly inefficiently but eco-effectively. A cherry tree produces thousands of blossoms and fruits without harming the environment. On the contrary, as soon as they fall to the earth they become nutrients for animals, plants, and the soil." A cherry tree doesn't save, deprive itself, or seek to avoid; it produces in abundance, but it produces only useful and beautiful things.

In Braungart's vision of the future, there are no more consumers, because things are no longer consumed in the usual sense. People can use products without any risk, because they are beneficial to the environment or endlessly recyclable. The previous concept of eco-efficiency, he says, can slow down the process of environmental pollution and the growing scarcity of raw materials, but not stop it. After all, what good is it if cars reduce their gasoline consumption by one half while the number of cars worldwide triples, or if we increase the percentage of synthetic materials that can be recycled, but these materials are afterward only suitable for products of lesser quality? We can't convert every plastic product into park benches, flower pots or noise barriers, or even burn them to generate energy, says Braungart. By contrast, eco-effective products would consist of only those plastics that can be reused without loss of quality, or their waste materials would be used to make new products. For instance, nitrogen oxides from automobile emissions could be converted to fertilizer.

What Does the Customer Want?

The fact that "cradle to cradle," abbreviated as C2C, is not an idle fantasy can be seen from the roughly 600 products that Braungart and his colleagues have now developed for and with companies around the world, including Airbus, BASF, Volkswagen, and Nike. For Nike, he designed fully recyclable shoes; for Trigema, a compostable T-shirt. In these products, only biodegradable materials are used, from the sewing thread to the dyes. The worn-out T-shirt can simply be thrown on the compost pile. Within half a year, fungi and bacteria will decompose the fibers without leaving any residue. Quite a lot can be improved in the case of packaging as well. Biodegradable materials are a particularly good choice here, because there is no reason why shampoo bottles, toothpaste tubes or yogurt cups have to outlive their contents by decades.

But how would complex products like automobiles or TVs be recycled? Braungart has a remedy on hand for this too: "We have to reinvent everything, beginning with the question 'What does the customer want?' What they don't want is to take on the responsibility for a TV with over 4,000 toxic chemicals; instead, they want is to see a movie. Instead of actually owning a car, they want to be mobile. In such cases, it makes sense to sell the utility rather than the product. It's a kind of ecological leasing." Braungart is thus suggesting that the manufacturer will retain ownership of the TV or car, and the customer will buy a service such as the right to watch television for five years, use a computer for three years, or drive 100,000 kilometers. After that the manufacturer will take the products back again.

Braungart is convinced this would have huge repercussions on the materials used. "Companies would then build not the cheapest thing but the best — the product that can best be reused. For example, in cooperation with an automaker we're building a car body whose components will be glued rather than welded together. After it's used, the body is immersed in a solution with bacteria that decompose the glue. The individual parts can then be reused," he says.

In another example, Braungart caused a sensation with the design of compostable seat cushions for the Airbus A380. "Products like this don't even have to be very expensive. Airbus actually cut costs by 20 percent with this product, because they didn't have to dispose of the old seat cushions as hazardous waste," he says.

C2C Nation

The C2C concept is attracting more and more fans around the world. In the U.S., for example, it is being promoted by Cameron Diaz, Brad Pitt, and Susan Sarandon. Steven Spielberg was so enthusiastic that he donated millions of dollars and wants to make a documentary about Braungart. In China and India, a special sort of ice cream packaging was launched. It remains a solid film when frozen but melts at room temperature. Since it also contains the seeds of rare plants, the film even contributes to biodiversity when it is thrown away. The Netherlands, meanwhile, have so much faith in Braungart's ideas that they "are well on their way to becoming a C2C nation, from daycare centers to the royal palace," he says. "We're working on new C2C ideas there in collaboration with construction and civil engineering companies and electronics manufacturers."

Only Germany remains somewhat skeptical of Braungart. "Many Germans romanticize nature too much, and they're quick to view technical and chemical innovations as a threat," he says. But some of the criticism is certainly of a substantive kind. For example, Friedrich Schmidt-Bleek, the long-standing director of the Wuppertal Institute for Climate, Environment and Energy and current president of the Factor 10 Institute in France, takes aim at the compostable seat cushions. "I can feel very comfortable on Michael's seat cushions. But I'm still waiting for detailed proposals on how to design the other 99.99 percent of the Airbus A380 according to his principles," he says. Schmidt-Bleek considers a perfectly circular recycling industry for all the products used by human beings to be "completely out of the question."

One thing is certain, however. The more C2C products there are in the future, the better it will be for the environment. Eco-effectiveness and eco-efficiency should not be seen as opposite poles in 2050. In fact, it's often faster and cheaper to reach the desired outcome when they are combined. On the one hand, we will be saving energy and resources and avoiding pollutants; on the other hand, we will be designing products for an intelligent C2C recycling industry whenever possible.

The rethinking has already begun. Many products are no longer judged solely according to price, performance, and quality. Today, a holistic point of view — a life-cycle assessment — is frequently a prerequisite for even taking part in the bidding for public works projects, for example when transit agencies order new trains. In such cases, the ecological compatibility of all stages of a product's life cycle —from the extraction and processing of raw materials to transport, usage, and disposal – is taken into account and evaluated (see p. 85). And in many cases ecology and economy are not contradictory. For example, a mere two-percent improvement in the energy consumption of a locomotive is enough to justify paying as much as ten percent more for it at the outset. Similar considerations apply to energy-saving household appliances and products. Buying appliances like these is a real win-win situation. They are good for the environment and for the buyer's wallet.

Like an Office Next Door – but Thousands of Miles Away

Ecological thinking and intelligent concepts like C2C, together with the massive use of simulation and automation technologies, will determine how competitive nations are in the future. This is the only way it will be possible to keep building factories in countries with high labor costs; those costs will then make up only a modest percentage of the value of a product. At the same time, however, the digitization of all production processes will help

Around the clock, and around the world. In the corporate network of the future, everyone will know immediately when something is changed in a product design, regardless of where in the world the change was made.

reinforce the process of globalization and lead to "around-the-clock production." Today, some global corporations have already established such close communication links between locations on different continents that the experts involved almost feel as though they are right next door to one another. In the case of software development, this sort of arrangement is especially easy to set up. Hundreds of programmers in the U.S., Europe, and India work together on large software packages, whether for medical imaging systems or new cell phone technologies.

Factories are now also using "dynamic 3-D data" that immediately show when a designer, production specialist or supplier has modified a detail, regardless of where in the world that person is working at the moment. In the past, design drawings were often sent back and forth via courier with handwritten comments, or they were scanned in and sent electronically. In the future, engineers will increasingly meet in video conferences that use virtual reality technologies. This way, designers can all simultaneously look at the same 3-D model of a vehicle or a turbine and make changes to it, and assembly specialists can quickly see whether components will fit together easily. The technologies that are being used today in just a few sectors, such as the automotive, aircraft, and power plant industries, will be commonplace throughout all industrial sectors in coming decades.

A Jam Session for Product Designers

Increasing global interconnectedness will bring with it completely new methods of product design, such as "crowd sourcing," in which consumers are asked directly to express their wishes and suggest improvements via the Internet. A number of companies are already trying out this approach. For example, the Web community submitted more than 170,000 ideas for the design of the new Fiat 500. This method also helped to create innovative LED lighting solutions from Osram and T-shirts made by the U.S. company Threadless, the first company to develop all of its products on the basis of proposals from customers. T-shirt enthusiasts submit approximately 1,000 ideas every week. If a design is printed, its creator receives $2,000 dollars and enjoys fame and glory in one of the world's most respected graphic design communities. In the future, many more companies will make use of the interactive elements of "Web 2.0." The objective is to spend less time and money on the development of new products and at the same time to create new products that better match customer expectations, whether it be the design of a new handbag, a vacuum cleaner, or the cockpit of the next generation of cars.

Likewise, in the core area of modern companies, the R&D department, the age of closed laboratory doors and brilliant solitary thinkers is over; the new motto is "open innovation." In addition to collaborating with universities and research institutes, companies are increasingly collaborating with other companies and organizing moderated discussion groups on the Internet involving thousands of experts. In an analogy to the creative jam sessions of jazz or blues musicians with their free interpretations, IBM has called these innovation sessions lasting several days "Innovation Jams." The largest one to date involved over 150,000 people from 67 companies in 104 countries. IBM reported that this session led to ten promising new business ideas that the company supported with 100 million dollars of seed money.

Open innovation has now progressed to the point where companies are hiring external service providers to apply their global networks to a research problem. Attractive sums of up to one million dollars are offered as remuneration. One of the best-known platforms is the Internet innovation marketplace "InnoCentive Challenge," which draws on over 200,000 specialists around the world from the fields of chemistry, physics, the life sciences, materials research, software, and agriculture. These virtual global networks of inventors and problem-solvers will be a part of everyday life in the year 2050.

Knowledge Workers in the World of Tomorrow

What does all this mean for the jobs of the future? There is no doubt that some work will continue to be performed by hand at manufacturing sites around the world in coming years, whether in assembly halls for automobiles and household appliances or in the sewing rooms of textile companies. However, more and more work processes will be taken over by automation systems and robots, especially when the processes can easily be standardized. But even if the factories in the industrial countries of tomorrow seem almost devoid of human beings, there will still be people in the background who monitor the processes and jump in if something goes wrong. "Increasingly, the domain of human work is not manufacturing itself but the assumption of controlling or monitoring duties such as quality assurance," says Günter Voß, who specializes in the sociology of industry and technology at the Chemnitz University of Technology in Germany. And we mustn't forget all the fields that require a great deal of creativity and social interaction, such as research and development, product design, strategy and organization, purchasing and marketing, consulting, sales, and service. Many people will still be needed in each of these areas.

But this is bad news for people with inadequate qualifications. In many cases, not even continuing education helps if a certain foundation is lacking. "Human capital can't be molded at will in just any old way; it's tied to living people with individual abilities and the paths they've taken in life," says Voß. "The crucial thing is to give people broad qualifications early on and in sufficient numbers." Knowledge and know-how are the drivers of progress today. One hundred years ago, about 80 percent of working people in the industrial countries toiled in fields or factories, and more than 50 percent still did so only 50 years ago. Today it is just the opposite. Only 20 percent still permorm classic laborer jobs, while 40 percent work in offices, the retail trade, or the service sectors, which include everything from bakers to plumbers. The remaining 40 percent can be called "knowledge workers." They work in research and development, management or consulting and in schools, universities or media companies. But there is scarcely any occupation that does not use advanced technology. Even an auto mechanic must now be able to use electronic test equipment, computer programs, and complex instruction manuals.

A Morning Lecture in Munich, Boston in the Afternoon

Since the knowledge needed in the professional world changes extremely rapidly, the employees of tomorrow will have to adjust to life-long learning. Universities and corporate training centers are already testing diverse forms of virtual universities and computer-assisted learning. In 2004, for example, a group of lectures took place simultaneously in Zurich, Munich, Warsaw, Beijing, and Tokyo; the organizers called this "experimental global instruction." Using video-conferencing lines, a professor projected

The seminar of the future. Whether live and in person or linked up via 3-D Internet, tomorrow's teaching will rely on multimedia experience.

his English-language presentation onto screens in different lecture halls, and students had an opportunity to ask questions at any time. They did not even have to be physically present in the lecture hall. All they had to do was to click on a Web page to follow the live lecture. Those who missed it could simply download a video complete with exercises.

These global Internet lectures provide a foretaste of how students will learn in the future: interactively with multimedia, independently of place and time, in a mixture of live, in-person events and electronic learning. The traditional lecture will not disappear, because the personal give-and-take among students and the guidance of tutors are extremely important — but it will be complemented by multimedia elements, particularly with regard to the subsequent work in teams spread throughout the world. Why shouldn't a future student in Munich attend a morning class at her own university, then take part in the afternoon Internet lecture of a Nobel laureate at the Massachusetts Institute of Technology in Boston, and get together with fellow students in California and Japan in the late evening to work on a joint research project in a virtual room with a video link-up and document sharing?

In the globalized world of 2050, there will be constant competition among not only companies but also educational institutions. Universities will develop new forms of cooperation and offer their instructional content worldwide. Education and the transfer of knowledge will become commodities. The students with the best chances in the labor market will be those who, in addition to having a good grasp of their disciplines, will have learned to adopt an open-minded, market-oriented perspective and to work in international teams. Women generally respond well to this sort of team-oriented, networked style of working. They will therefore find a great many spheres of activity in the working world of tomorrow and rise to leadership positions to a much greater extent than today, especially since the trend toward telecommuting and home offices is making it easier to combine work and family.

In research and development, the people who enjoy the most success will be those who can see beyond the horizons of their respective specialized fields, because many of the innovations of tomorrow will arise at the meeting points of various disciplines. For example, those who hope to build a lab for medical analysis on a microchip need to know a whole lot about microcircuitry, but they also need knowledge from the fields of chemistry, biology, physics, sensor technology, and data processing. The same applies to decentralized supplies of energy. In this case, it is important to understand not only wind power stations and solar cells but also industrial control systems, energy storage devices, and communications systems.

But all of this is certainly not just about technology. For example, the success of the iPod, iPad, and iPhone is ultimately based not on technical innovations but on the fact that Steve Jobs and his team at Apple succeeded in combining known solutions for wireless communications and computer technology with a clear and consistent design, a simple and intuitive user interface, a massive marketing campaign, and a large fan community. And don't forget the partnerships with independent programmers who design hundreds of thousands of applications, or "apps," that offer everything from brain exercises to navigation systems and the latest recipes for the kitchen. Apple combined all of that with the appropriate marketing, sales, and business ideas for music, games, programs, and films. This is how otherwise ordinary things are turned into something completely novel: in this case, an ultra-popular digital lifestyle embraced by millions of people.

□ *Cairo, March 2050. The "Center for Living Memory" is opened to visitors for the first time. At the entrance, guests are issued ultra-thin datagloves, 3-D glasses, small headphones, and flat navigation devices in a classic iPad design. All of these devices are wirelessly linked to the museum's database and interactive exhibits. For example, if a visitor enters the digital aura of the Hall of the 19th Egyptian dynasty, he or she will see many things from around 1290 B.C. — thanks to the 3-D glasses. There will be ointment jars, sculptures of Horus and Anubis, and artfully crafted hieroglyphics displayed on walls. However, except for a sarcophagus and the mummy of the favorite wife of Ramses II, the burial chamber actually contains very few real objects. It's the datagloves that enable visitors to touch non-existent vases and see what's inside. All the devices follow visitor movements in realtime. What's more, museum guests can also converse with a virtual queen who rises up from the sarcophagus. "I am Nefertari, beloved of the goddess Mut, Great King's wife, hereditary Princess of the Two Lands, most beautiful of them," she says gracefully. She then patiently answers questions about life as it was 3,300 years ago, addressing topics ranging from religion to details of how her clothing was woven. Sometimes she even lets visitors touch her fine fabrics with their gloves...*

3-D GAMES AND REAL LIFE

When people in 2050 look back on the first decade of the 21st century, they may conclude that this was the era in which two great revolutions were launched. The first was ushered in when the human race finally began to understand that it needed to radically restructure its energy system; the second started with the creation of the worldwide social networks. These networks were not established by government agencies or major organizations but instead by hundreds of millions of private individuals. They took the form of online communities like Facebook and Myspace, platforms for exchanging images and videos like Flickr and Youtube, marketplaces like eBay, online journals in the form of blogs and Twitter messages, and fantasy role playing games like *World of Warcraft* that allowed thousands of users from around the world to participate simultaneously.

Many people today spend a lot of their free time with such social networks. They participate in them for free, and some individuals even display great dedication — such as those in the Wiki Communities who post content on the Internet and make it available to all other users. The most well known of these is Wikipedia, whose founder, Jimmy Wales, decided in 2001 to do nothing less than "put the entirety of human knowledge" on display. Wikipedia now offers information in more than 260 languages, with more than one million registered users providing content to help make Wales' vision a reality.

All of this has occurred within just a single decade. Facebook, for example, has only been around since 2004 — but six years later it already had 500 million users. The World Wide Web of online links was invented by Tim Berners-Lee from the CERN particle accelerator facility in 1989, the year the Berlin Wall came down. Berners-Lee's intention at the time was to facilitate the exchange of research results with colleagues in the science community. Just six years after that, 40 million people already had access to the Internet; by 2007, the figure had reached 1.2 billion and it will likely quadruple before the end of the current decade.

This rapid expansion also has to do with the fact that more than five billion people will have cell phones by 2015, many of which will be smart-phones that can connect to the Internet. The age of cell phones also didn't begin until the 1990s. Now, some regions in Africa and Asia actually no longer bother to set up fixed-line connections anymore, in line with the motto "From no phone to mobile phone." There were 720 million cells phones operating worldwide in 2000 — but today in China alone there are already 850 million wireless users, and nearly five billion around the world.

Such rapid growth makes it difficult to take seriously forecasts regarding computer and communication technology in 2050, just as no one could predict in 1970 what the state of technology would be in 2010. Indeed, there were only 11,000 bulky and expensive mobile phones in all of Germany in 1970, while the term Internet wouldn't be coined until 12 years later. PCs were unknown as well at that time. Consider the following: The onboard computer used in the Apollo 11 space craft had an internal memory of 33 kilobits. Most of today's laptops can store one million times that amount of data.

Although many of their details have changed, houses, cars, and power plants still look pretty much the way they did in 1970 — but information and communication technology is completely different. That's not surprising, given that a new house or power plant is built only once every couple of decades, and most cars have a lifespan of more than ten years. This means that a period of 40 years corresponds to one, or a maximum of four, product generations. With computers and cell phones, on the other hand, it's at least ten to 15 generations. It's impossible to make predictions over such a long period of time when developments proceed so rapidly. In terms of a human generation of 25 years, that would be like someone asking Voltaire in 1750 what the world would be like in 2050.

A Thousand-Fold Increase Every 20 Years

There are, however, certain general principles that can be applied when looking into the future. One of these is Moore's Law, which was formulated in 1965 by Gordon Moore, one of the co-founders of computer chip manufacturer Intel. This law states that the number of transistors that can be placed on a fingernail-sized integrated circuit will double every 18 to 24 months. This number has a huge effect on computer power and chip storage capacity. Moore's Law has remained valid since the first microprocessor was marketed in 1971 as the "Intel 4004." That unit had 2300 transistors — and sure enough, 20 years later the Intel 80486 boasted 1.2 million transistors, while 20 years after that the IBM Power7 was fitted with 1.2 billion. That corresponds to a thousand-fold increase within 20 years — without any significant increase in cost for microchips.

Land of the Lilliputians. Electronic components are becoming smaller and smaller. Shown here is a miniature pressure and temperature sensor as compared to the dimensions of an ant.

The storage capacity of magnetic hard drives has also risen more than a thousand-fold over the last 20 years. Whereas in 1990, they could store up to a gigabyte, their maximum capacity today is two terabytes. One gigabyte corresponds to the text content of a large cabinet full of books; a terabyte is like a library with a million books. And just as USB flash drives are increasingly replacing CDs and DVDs as a storage medium, so too will hard disk drives be unable to keep up in a future in which storage cells made of transistors become so small that trillions of them will fit on a chip — possibly stacked three dimensionally one on top of the other.

Such storage density might even reach the level of holographic storage units that use lasers to "write" data onto crystals or plastic. Such materials could store around two terabytes per cubic centimeter. The only thing that could top that is the storage medium that each life form has in its cells: DNA. The entire human genome, with its three billion letters, fits into a cube with an edge length of just a little more than a micrometer. That corresponds to an unbelievable storage capacity of 200 million terabytes per cubic centimeter. However, technically speaking, building such a DNA storage unit is virtually impossible because the read-out speed in nature is very low — only 10 to 20 bytes per second, as opposed to USB flash drives that transfer data at millions of times that speed.

If scientists hope to pack more and more transistors onto a silicon chip, they will have to make individual circuit elements smaller and smaller. The finest structures on these chips currently have a size of approximately 50 nanometers, or one thousandth the diameter of a human hair. A single transistor is therefore smaller than a flu virus, but researchers are already testing methods for shrinking its dimensions by a factor of five to ten. Most microelectronics experts believe that Moore's Law will reach a limit to its validity between 2020 and 2030 because circuit elements will then be reduced to the size of just a few atoms and it will become extremely difficult to dissipate the heat created during computing. However, this would still mean that computing power would have further increased 500-fold by 2030 — and scientists will certainly not have exhausted all their ideas by that time.

Quantum Computers Think Differently

In principle, a single electron or atom would be enough to store one data element (bit). The rules for computers would then change, however, because such tiny systems would be governed by the laws of quantum physics. These systems would no longer be in a clearly defined state but instead in a type of simultaneous overlay of all possible states. A conventional computer calculates with bits, which have a value of 0 or 1, but a bit in a quantum computer can with a certain probability be 0 and 1 at the same time. In this situation, it would no longer be possible to conduct calculations in a conventional manner — but it would be possible, if cleverly done, to process thousands or even millions of bits of data simultaneously.

Encrypted codes could then be cracked quickly through a massive parallel hacker attack. Moreover, if scientists succeed in coupling a quantum computer with a system that can learn from examples what certain things look like (neural networks), they would be able to create an extremely efficient pattern recognition technique that could easily identify everything from addresses scribbled on paper to human faces. It could also be used to unmask computer viruses by comparing the content of a suspicious file with the symbol sequences of innumerable known types of malicious software. There are no such powerful quantum computers today, but scientists around the world are working feverishly to create them.

Movies from Wireless Networks

Just as we've seen huge increases in data processing and storage capacity, we've also seen thousand-fold increases in data transfer rates over the last decades. Whereas ten years ago, customers were happy to have an ISDN modem with a data transfer rate of 64 kilobits per second, they now demand ADSL or VDSL systems that enable them to download 100 to 1,000 times more data from the Internet during the same time. Japan already has companies that offer households glass-fiber connections with a maximum transfer rate of one gigabit per second — nearly 16,000 times faster than the old ISDN links. In addition, the connection backbones between large servers are set to be upgraded to 100 gigabits per second, and tests are already under way with systems that deliver 1,000 gigabits per second.

All of this is becoming necessary because huge amounts of data now need to be transferred to run videos, for example. Simultaneous work on 3-D product simulations, the systems for which are now being developed, also requires high data transfer rates. Good Internet radio reception requires a rate of only a few hundred kilobits per second, while watching a video in DVD quality translates into a couple of megabits. But staging a video conference with near-real images of participants and high-resolution 3-D product data (including real-time processing) would necessitate several gigabits per second.

In the future, it will be possible to carry out such activities even via cell phone, as the data transfer rates for these devices are increasing more than one hundred fold each decade. The users of the first GSM cell phones in the 1990s had only 9.6 kilobits per second available to them, which means basically all they could do was make calls and send text messages. But the GPRS and EDGE data transfer standards introduced a few years ago deliver rates of 100 to 300 kilobits per second, which enable Internet surfing and transmission of small video files. UMTS networks with High-Speed Downlink Packet Access (HSDPA) are even more convenient, as they make possible transfer rates of 3.6 to 7.2 megabits per second, and in some cases rates that are even four times faster. Some one billion people around the world are expected to be enjoying such data transfer rates by 2012.

And that's certainly not the end of the story. Wireless local area networks (WLAN) already provide data transfer rates of 54 megabits per second and will soon become six to ten times faster. However, receivers of such data usually have to be located within a maximum radius of one hundred meters of the transmitter, while the cell in the classic wireless network has a range of several kilometers. The successor to UMTS, LTE, which is currently being built, will likely have radio cells spread less than a kilometer apart, which means more transceivers will need to be set up. LTE will be able to achieve data transfer rates of 100 to 300 megabits per second. Signals have already been sent in labs at a speed of one gigabit per second, which means movie fans should be able to watch a 3-D film in realtime with no problems around the middle of the 2020s. They will also be able to download cinema-quality resolution from the Internet to their cell phones while sitting on a beach, for example. The only question is what kind of display they will watch it on.

Avatars Shopping in Virtual Stores

While large, flat, and even bendable high-resolution displays will be omnipresent in the buildings, vehicles, and cities of tomorrow, a large display is no good for portable devices — unless, of course, it can be folded or rolled out (see p. 204). Much easier to develop are cell phones with integrated LED or laser projectors that can show images and films on the nearest wall. The first such devices do in fact exist.

Web surfers on a beach in the future might also use glasses fitted with two small projectors that precisely focus images directly into their eyes. They would then see a three-dimensional object that "floats" right in front of them. A dataglove would allow them to grab, move, and turn the object, and tiny pressure transmitters or targeted vibrations in the glove would give users a feeling for the object's texture.

This would make it possible, for example, to go shopping at a virtual 3-D Internet store without ever leaving the beach. Vacationers could not only view from all sides the robot toy they're thinking of buying, but also "touch" it. The great thing for other people on the beach is that no one would be bothered by this. Anyone watching would only see someone with glasses and a glove groping around in the air for no apparent reason. Life in the future therefore also promises to be amusing ... especially since remote shoppers won't be limited to a visit to a virtual department store. That's because real stores, to the extent that they'll still exist, will also be able to offer remote visits via robots. In this case, a shopper would rent a real robot on the Internet and then use its camera eyes, arms, hands, and legs to stroll through the store and examine merchandise — a kind of metal avatar, so to speak.

And speaking of avatars, three-dimensional cinema achieved its true breakthrough with the release in 2010 of James Cameron's film *Avatar*, which became the most successful blockbuster of all time. It's therefore not surprising that television manufacturers now believe the time has come for 3-D sets. Three-dimensional television isn't that difficult from a technical point of view. All that needs to be done is to film scenes the way people see them — in other words, with two cameras set apart at the same distance as our two eyes, which also means the left camera records images from a slightly different angle than the right one. The television image can then be transmitted in double, whereby the viewer's left eye should only see images recorded with the left camera, and vice versa.

This can be done with polarized glasses like those used in the cinema. These glasses let light with a specific plane of oscillation pass through, while keeping light oscillating in another plane out. Alternatively, electronic shutter glasses can be used that alternately close the lens over the right or left eye for a few milliseconds, while the 3-D television synchronously shows the proper image for the lens that's open at a given moment. These shifts are so rapid that the brain believes the images are appearing simultaneously, creating a 3-D effect.

3-D TV without Glasses

The Fraunhofer Heinrich Hertz Institute in Berlin has developed a solution that requires no glasses at all for the viewer: autostereoscopic displays. They have a plate made out of rod-shaped lenses or prisms mounted over their screens. These direct the light of one column with pixels to one eye, while also channeling the light from the adjacent column to the other. Packing such lenticular lenses as closely together as possible allows several individuals to experience the 3-D effect simultaneously. "Here, however, you won't just need two but instead at least ten different views of a scene — even better would be 30 or 50 views with large displays, which of course means the required data transfer rate would be higher," says Thomas Wiegand, head of the Image Processing department at the Heinrich Hertz Institute.

Devices with simple autostereoscopic displays are already being used in industry and hospitals to view new products or images of the inside of the body in 3-D. The Berlin-based institute has also developed software for controlling the devices through hand gestures recorded in realtime by two infrared cameras mounted on the display. The software interprets hand movements and translates them into computer commands for turning or tipping objects, for example. No matter which technology eventually becomes the standard, Wiegand is sure of one thing: "In 20 years, we'll be looking back at two-dimensional television the same way we now look back on black-and-white movies."

Still, 3-D films aren't suitable for all subjects. After all, a blockbuster like *Avatar* was worked on for many years to achieve the perfect spatial effect, but that type of effort is too expensive for most movies — not to mention talk shows or soap operas. Experts therefore believe the greatest appeal lies in sporting events like soccer games, which can be made even more exceptional in 3-D. The sex-film industry is also looking forward to new developments in 3-D TV; that sector has always been among the first to take advantage of new media technology for photography, film, video, and the Internet. In addition, experts view the development of 3-D games as a "killer application" because the games industry operates according to the principle that the more realistic a fantasy world is and the more deeply players can immerse themselves in it, the more successful a game will be.

Ride the Magic Carpet

The Japanese company Nintendo has already taken major steps toward better human-machine interaction with its Wii remote control, whose integrated acceleration sensors enable it to react directly to arm movements, and its Balance Board. Players stand on the latter and then shift their bodies in order to send signals to the game console. The 3-D virtual reality systems of the future will go far beyond that, however. A taste of what's to come can be experienced by visitors to the Cyberneum operated by the Max Planck Institute for Biological Cybernetics in Tübingen, Germany.

In 2008, the Cyberneum became "the world's first walkway through a virtual word in which free movement in all directions is possible," according to Heinrich Bülthoff, a director at the institute. Bülthoff's "test subjects" move along a kind of carpet consisting of conveyor belts so cleverly controlled that the people on them are always near the center of a 4x4-meter surface. If they move forward, the belt carries them back; if they turn left, the belt moves to the right; and if they turn around, the belt reverses its direction.

The video helmets the visitors wear allow them to stroll through the ancient city of Pompeii, for example, and see it as it may have looked before being destroyed by a volcanic eruption in 79 A.D. — with its white plastered houses, red tiled roofs, temples, theaters, baths, murals, and statues. Max Planck researchers are not looking to develop a new theme park attraction here, of course. "We are examining the interaction between perception and movement," Bülthoff explains. "What do our senses do, how do they work together, which types of stimuli does the brain react to, and how does it make decisions?"

Those who enjoy computer games may one day find more simplified versions of the Max Planck conveyor belts in their living rooms, from which they'll be able to explore fantasy worlds like *World of Warcraft*. Equipped with data glasses and a remote control console with motion sensors that can be used as a sword or magic wand, they will be able to transform themselves into the knights, elves, witches, or wizards they always wanted to be.

Their fellow players with whom they experience their adventures in this world will be people like themselves who are at the same time also standing on a "magic carpet" somewhere else in the world. All will be linked together via the Internet and fast data connections. It's even possible that

Virtual visit to Pompeii. Visitors to the Cyberneum in Tübingen can stroll through virtual worlds and feel as if they're actually there, thanks to a video helmet and a smart walkway control system.

some will be standing in a cubicle that generates wind and vibrations, as well as realistic noises and scents.

This would definitely be the ultimate experience for a gamer. What psychologists might think of it is another story, however. Indeed, some people's involvement in fantasy worlds already goes too far. On the one hand, it's true that such games require players to utilize much more of their intellectual abilities then they do when they simply sit and watch TV. They need to display sensitivity, respond to events quickly, think strategically, and work in teams. When players are successful, it also boosts their self-confidence. On the other hand, some people develop emotional attachments to a world that's not real — and once such people become deeply involved in the game, it's hard for them to stop. The approximately 12 million players of *World of Warcraft* tell a clear story here.

In addition, the peer pressure involved in solving a task in a team can get so heavy that the game can govern a person's entire daily routine because all team members have to play at the same time. If work, school, and real friends are then neglected, the virtual world becomes more important than the real one. One study has shown, for example that 15-year-olds who play *World of Warcraft* spend an average of nearly fours hours a day at their computers.

The study, which was conducted in 2008 by the Criminological Research Institute of Lower Saxony, concluded that out of all ninth-grade students in Germany, more than 14,000 are addicted to computer games, and an additional 23,000 are in danger of becoming addicted. Researchers and parents are therefore worried about what will happen when the immersion experience in virtual worlds becomes even more intense than it is today.

Thought-Controlled Flippers

Some scientists, like U.S. researcher Ray Kurzweil and astrophysicist Stephen Hawking, think even further ahead and postulate the development of high-powered computers on the micro and nano scale that could be directly hooked up to the human brain. Such a computer would greatly expand the brain's horizons and also enable it to be connected to other brains, and even to a "world computer" via mobile communication systems. In such a future, there would no longer be any difference between the virtual and real worlds. Taken to the extreme, it could mean that the human mind might eventually be released from its body and become part of a "global mind."

To most people, this sounds like the movie *The Matrix* come to life. Nevertheless, researchers working under Peter Fromherz at the Max Planck Institute for Biochemistry in Martinsried near Munich demonstrated in the 1990s that nerve cells could be placed on a silicon chip and communicate with it. The chip uses electrical impulses to stimulate the cells, which forward the signals to other nerve cells via their biological contact points, the synapses. The activity of the latter then triggers a change in the voltage of the lower lying transistor, which further processes the signal. In other words, the nerve cells and the chip communicate with one another — and without any damage to either.

This phenomenon is very important for fundamental research. And while no serious scientist is currently considering the implantation of silicon chips in a human brain, such a drastic step might not be necessary, because virtually any level of empowerment would be a huge step forward if, for example, it enabled paraplegics or people in a vegetative state to use their brain power to induce some kind of action. To accomplish this, Klaus-Robert Müller, a professor at Berlin Technical University, is using a cap containing electrodes attached to the scalp that enable a rough depiction of the brain's electric activity. Such a device can be used to create a thought-control system so fast as to enable users to operate pinball flippers with their thoughts.

Müller's test subjects sit motionless in front of a pinball machine and operate the flippers solely by imagining their movements — not by carrying them out. With just a little practice, anyone can learn to do this within 30 minutes, says Müller. The computer that is trained to evaluate brain electrical activity during this time can later reliably recognize whether the pinball player is thinking about hitting the ball with the left or right flipper at any given moment. The reaction times achieved here are significantly lower than one second. It's not possible to read more complex thoughts, of course, but an advanced version of this system could enable paralyzed individuals to operate wheelchairs through thoughts, or perhaps even write text on a computer in the same manner.

Invisible Graffiti and Virtual Pharaohs

Controlling devices with thoughts, exploring virtual worlds — all of this sounds as if an invisible parallel world will exist in the future. And this is exactly what will happen. Researchers at Johannes Kepler University in Linz, Austria, for example, are currently developing a new information system that works with invisible labels. They call their new system "Digital Graffiti," and in the future all of the university's 10,000 students and 2,000 teaching and administrative staff will be able to use it. The basic idea is simple. Students and staff can use their cell phones to post virtual messages at any location for either specific addressees or for anyone. When a person authorized to receive the message approaches this location, the message is automatically sent from a server where it's been stored to the user's cell phone, thus making the graffiti visible, so to speak.

All types of messages can be left with the system, including information on which room a lecture is to be held in, or descriptions of interesting sights on campus. The graffiti messages can also be date-stamped, ensuring that they are only available for a certain period of time, after which they disappear. This type of location-based service will be very common in the future, especially since it will also soon be possible to pinpoint cell phones down to the last meter. Pedestrian navigation systems designed on this basis will become very popular and very much in demand, as they will allow people in unfamiliar cities to quickly locate interesting sights, restaurants, or the nearest bus stop – including information on when the next bus is departing.

Another interesting development for many will be the expansion of social networks into the real world through "friend-finder" software that bears an amazing resemblance to the marauder's map seen in *Harry Potter*. The program displays on a cell phone navigation map the locations of friends in the vicinity, provided they have activated this function themselves. Google Maps already offers a simple location-finding service called "Latitude," which allows each user to define how precisely their current location should be transmitted – in other words, their approximate position, only the city they're in, or absolutely no information should they wish to retain their privacy.

Location-based services linked to other types of data are particularly useful in museums. Here, visitors could receive interesting information about certain exhibits either as digital graffiti on their cell phones or via augmented reality software. A person could point his or her cell phone at a sarcophagus, after which he or she would not only see a picture of the actual object on their display but also a lot of additional, superimposed information. The sarcophagus could be made transparent, for example, revealing a view of the mummy and grave goods inside.

Taking the concept a little further, visitors to a museum could rent or purchase data glasses, earphones and a small computer tablet (like an iPad) at the ticket counter. With the glasses, they would not only see the sarcophagus but also perhaps watch Nefertari, the favorite wife of Ramses II, as she rises from the coffin and talks about ancient Egypt. They could even use the tablet to contact the virtual queen and ask her questions about the sculptures or murals that appear only virtually in the otherwise barren room (see p. 178).

Data Mining in the Cloud

Whether at a museum, on the move, in an office, or at home — connections to the all-encompassing data network will be as normal a part of everyday life in the future as having electricity is today. "Always on" will be the motto here — a world in which anyone can access the invisible information of the Internet at any time without having to think at all about the technology behind it. Rapid transfer of huge amounts of data will be a matter of course, as will the circumstance that various devices — from home PCs and smart phones to vehicle navigation systems — will be able to communicate with and understand one another.

These devices will also be able to access Web resources independently. "Cloud computing" is how experts refer to this phenomenon in which a person carries out a task using memory, computing power, and software resources located somewhere else on the Web. Cloud computing allows thousands of computers around the world to be linked together in order to solve particle physics or climate research problems, for example, or help with the search for new medications or even extraterrestrial life.

However, as the flood of data swells, it becomes increasingly important to pin down how truly relevant information can be found. That's why data mining is set to become an industry in its own right in the future. Sophisticated programs are already working on tracking down criminals who use stolen credit cards. These programs analyze innumerable card transactions to generate a pattern of how normal customers behave. They then isolate suspicious behavior, such as when a credit card is used to withdraw money from ATMs at short intervals, or when a customer suddenly begins gambling in Internet casinos, despite the fact that he or she has never done so before. The software then warns the credit card company that the card in question may have been stolen.

Software Agents as Virtual Servants

Similar programs use travel routes, visited websites, and bank transfers in an attempt to track down terrorists. Others classify customers according to specific purchaser groups and offer online shopping tips. Some are even capable of learning, such as the Pandora Internet radio system, which allows listeners to initially train it by providing it with feedback on whether or not they like the songs it plays for them. After this training phase, the software suggests other tunes, most of which, amazingly enough, correspond to the listener's tastes and are songs they never would have thought of asking for themselves. Those who wish to spare themselves the effort — and pleasure — of searching for TV shows themselves will be able to purchase an adaptive television set that combs the Internet or TV stations for programs that might be of interest. This type of software is extremely valuable to advertising firms as well because it allows them to send their commercials directly to those who might want to buy the associated products.

Adaptive software also forms the basis for software agents — autonomously operating self-contained programs that act in the virtual world like robots do in the real one. Such agents are already used today in computer games and to search through large databases. Among other things, researchers use them to simulate the behavior of large numbers of people, and to analyze transaction patterns on electronic stock-trading platforms. In the future, they might also be used to plan complete travel itineraries, book flights, reserve hotel rooms, and reduce the work load in administrative offices.

But computers still operate in terms of zeros and ones. And although they are outstanding calculation wizards, they are rather poor at interpreting images and texts, making associations, and identifying connections. In other words, they don't "understand" the world very well. They're able to beat a world chess champion because chess is based on clear rules, but they haven't advanced very far in translating from one language to another, as that skill requires a lot of everyday knowledge. For example, the sentence "the bank has collapsed" could either mean that a bank building's structure was unable to withstand an earthquake, or that a financial institution has failed. A computer needs to know all these nuances and possible contexts if it's to be able to translate text correctly. That's why dictation and translation programs are useful only for very limited applications, such as legal or medical texts, but aren't much help with translating colloquial language.

When Robots Weep

Human beings are guided through their everyday lives primarily by their experience and intuition, which enable them to quickly make an emotional assessment of a given situation. Our feelings allow us to focus on just a few of the huge number of possible decisions we can make and filter out the rest. Feelings also often determine which things make sense for us to remember, and what might better be forgotten. Today's computers are already capable of interpreting human gestures and facial expressions by comparing them with learned images. They can thus recognize the six basic feelings of anger, sadness, joy, fear, disgust, and surprise — but obviously that doesn't mean they can feel emotions themselves.

Electronic butlers. Modern robots are already capable of flawlessly serving a cup of tea without spilling a drop.

No one knows how intuition and feelings can be transferred onto electronic circuits. Doing so might possibly require recreating the complex biochemical environment of life forms, with all their hormones, neurotransmitters, and endorphins. Creating emotional intelligence might also necessitate establishing a social structure and somehow inducing sensitivity and sympathy for other creatures. However, it might also be sufficient to link adaptive electronic circuits with a built-in system of rewards and penalties — in other words, rewarding a robot or computer for successfully completing an action, while also deterring it from undesirable behavior.

Markus Schneider, a student at Ravensburg-Weingarten College in Germany, implemented such a system in a simplified form when he developed a four-legged robot named "Thekla" for his Master's thesis. Schneider also received Germany's Computer Science Award in 2008 for his work. Thekla learns how to walk not with a built-in program but instead through rewards and punishment. The robot begins by making random movements and then receives positive feedback when it succeeds in using its legs and joints in such a manner that it moves forward.

The advantage of this method over those with set rules is that it also works with poor surfaces. Here, Thekla changes its movement patterns in line with the condition of the surface beneath it, which is why it walks just as well on smooth floors as it does on ground covered with debris. That would be ideal for a robot that might have to operate independently in an alien environment, such as the surface of Mars.

If researchers could also imbue a robot with curiosity about the world around it, they would come quite close to creating an independent being. The most ambitious scientists in this field are to be found at the Massachusetts Institute of Technology's Media Lab in Boston. Cynthia Breazeal heads the working group for personal robots at the lab and is viewed around the world as a pioneer in the field of sociable robots and human-machine interaction. Breazeal's most advanced sociable robot is called Leonardo – a cute creature around 75 centimeters high with furry ears, stuffed animal eyes, and childlike hands.

The little elf is curious about its environment, likes to try things out, and learns by imitating humans and their gestures and facial expressions. When Leonardo recognizes a face, it follows its movements with its head in an attempt to make eye contact. It registers sounds and analyzes the tone and intensity of voices in order to coordinate its own emotional reactions with the impressions it gains. It also tries to recognize how humans respond to its own facial reactions so that it can learn for the future.

A robot like Leonardo only mimics feelings, of course, but it does this so perfectly that it elicits real feelings from the people who meet it, who then want to play with it, challenge it, teach it tricks like a pet, and pick it up and hold it. To acquire everyday knowledge, future robots will need to do more than just stand on a table or explore a lab, as Leonardo does. They will also have to go out into the real world, grow up there, and learn new things as a child does. As the performance of microchips once again increases a thousand-fold over the next few decades, scientists will definitely carry out such experiments – and we can look forward to the results with great anticipation.

An Internet of Things: Talking Posters and Radio-Enabled Furniture

During the next few decades, the Internet will evolve from its current form as a giant data collector (Web 1.0) and platform for social interaction (Web 2.0). Researchers are already working on Web 3.0, which will also include semantic features — in other words, the ability to understand the meaning of words and images. Future computers will be able to understand the connections between different terms, the structure of sentences, and the meaning of images. Such a computer could be quizzed using colloquial language, or could independently provide a physician with comparative images for a specific X-ray, treatment data on patients with similar conditions, and relevant reports from medical publications. This type of digital doctor's assistant would make diagnoses and treatment decisions much easier.

The thousand-fold increase in microchip performance expected over the next decades also means that over the long term, the price per memory unit or computing operation will decrease by the same factor. As a result, microchips will become so small and cheap that they could be installed in practically any kind of everyday object. Myriads of tiny brains with sensors and wireless data transfer components could then be built into lamps, furniture, clothing, and even medical implants, where they would search the blood for cancer cells or ensure proper dosages of medications.

Scientists therefore predict an "Internet of things," consisting of objects that know their location, can communicate and, if necessary, organize themselves. For example, airplanes could drop sensors over forests like confetti in order to identify avalanches, forest fires, or pollutants at the earliest possible moment. They would then pass this information on among themselves until it eventually reaches a control center that notifies human technicians.

The application possibilities for such an Internet of things are as varied as the human imagination. Movie posters could be fitted with chips that show the observer's smart phone how to get to the nearest cinema where the film is playing; shirts could tell washing machines the temperature they need to be washed at; and cell phones might inform their user in a café that the woman at the next table is looking to sell her car. This would seem like some kind of black magic to people from past ages — but by 2050, the Internet of things and virtual parallel worlds will make them a matter of course.

There are limits to this as well, however. For example, some experts believe the refrigerator that orders food by itself — so often described in future scenarios — will never be built because no one will want their refrigerator to make food-shopping decisions for them. Object localization via the Internet is also a double-edged sword. On the one hand, it would be useful to someone who has misplaced his or her keys or wallet, but it would also be more than dubious if a government used such a function to monitor its citizens (see p. 150). Discussions will have to be held over and over again in the future regarding which technical solutions actually make things more convenient and safe and which are simply unacceptable because they excessively intrude on privacy and individual freedom.

Talking objects. In the future Internet of things, posters will communicate with the cell phones of passersby, providing them with information on lectures, performances, etc. that the cell phone user might be interested in.

For the first time in history, the Internet has given the human race a tool that can be used to build a truly global community with direct access to a huge pool of knowledge and the opportunity for citizens to directly participate in political decisions. Indeed, Internet voting offers a way to implement direct democracy. The first steps in this direction have already been taken. Countries such as Canada and Singapore are gradually transferring all citizen-government interactions to the Web — everything from applying for commercial permits to submitting tax returns. Italy has introduced a smart card that serves as both an ID and health insurance card and can be used for submitting digital tax returns, making electronic payments, and voting. But real Internet elections have only been carried out in pilot projects. Some of these —such as those in the UK and U.S. — were then discontinued due to insufficient security. Nevertheless, the further development of encryption mechanisms and digital signatures will likely make Internet elections a reality by 2050.

Evolving Aspirations and Invisible Technology

What will all these developments mean for the society of the future? "Modern man rushes through the technologically advanced city, dominated by products of every kind, always accessible, yet somehow absent. Oppressed by the rapid transformations of his external and internal impressions, he seeks his salvation in nervous superficiality." These words were written by the German philosopher Georg Simmel about life in Berlin — in 1900. Simmel claimed the urbanite would become lonely because the only way to master the constant flow of stimuli would be to deaden his feelings. Things haven't gone that far.

Quite the contrary, as Heinz Bude, a professor at the University of Kassel and an expert on contemporary societies, says, "the future will be much less dramatic than people often claim. Our views about the influence of technology are exaggerated. There's no doubt that new technologies encourage more social contacts, but they also allow people to communicate for more trivial reasons."

This view is confirmed by the mostly irrelevant Twitter tweets and the common Facebook question: "What are you up to?" Some 250 Facebook friends — but none in real life when you need help or a shoulder to lean on — is this what the future will be like? No, says William J. Mitchell, a Professor of Architecture and Media Arts and Sciences at the Massachusetts Institute of Technology in Boston. "Personal communication has always been a rare and valuable thing. If you ask people in a survey the most common reason why electronic messages are sent, the answer they'll give you is to arrange a meet-up," he says.

Technology, Mitchell says, has changed the way people communicate and when they communicate, but people themselves don't change that much. "Sitting in a café is one of life's pleasant experiences — and that's something that will hardly change," says Mitchell. Moreover, the cities of tomorrow won't look like something out of Metropolis or Blade Runner. "In a paradox, the city of the 21st century is not going to look as high-tech as the city of the 20th century," says Mitchell, who instead is convinced that microchips and miniaturization will generate the opposite effect. "Now that technology is getting smaller and more robust, it is disappearing into your pocket, and into the woodwork, enabling the use of space to become more flexible. A telephone used to be part of the architecture, attached to the wall. Now that the phone is in your pocket," he observes. Such a development offers an advantage in that rooms will no longer have to be built around technical equipment. "That frees up architects to go back to designing spaces around very fundamental human requirements—sociability, light, air, view—the basic needs and pleasures of life," he adds.

Many researchers also believe that every development generates a countermovement. People in the future will therefore not only want to travel through virtual worlds; they will also feel a strong urge to experience real life through encounters with other cultures, excursions in jungles, visits to the Amazon to see howler monkeys and red river dolphins, trips to whatever coral reefs still exist, and journeys to view animals in their natural habitats – animals that are perhaps the last of their species or have been set off into the wild and now reside in protected national parks.

"Travel is the yearning for life," Kurt Tucholsky, an early twentieth century German writer and journalist, once said. People are constantly searching for new emotions and experiences, interesting stories, and the fulfillment of dreams – at least that's what the tourism experts say. However, the way they conduct this search depends on which values are important to them.

Back at the turn of the millennium, lifestyle researchers identified a new type of consumer: LOHAS, or those with a "Lifestyle of Health and Sustainability" and who therefore place great value on health and environmental consciousness when choosing foods, clothing, apartments and houses, and vacation spots. It's quite possible that these researchers will be proven correct in their belief that the LOHAS attitude will attract more and more adherents. That, in turn, would fit in perfectly with the sixth Kondratiev Wave – the beginning of a new age that focuses on health and the environment (see p. 12).

Still, the traditional urge for adventure will live on, and humans will continue to push the limits of what they can do and where they can go. Even today, some companies are already offering trips to the deep sea in special submarines – and trips into space as well. Firms like Astrium Space Transportation and Virgin Galactic plan to market the latter, which are essentially two-hour journeys to an altitude of 100 kilometers, at the start of 2012. It can be expected that this will be followed in the next few decades by orbital hotels, and at some point flights to the moon — or even Mars.

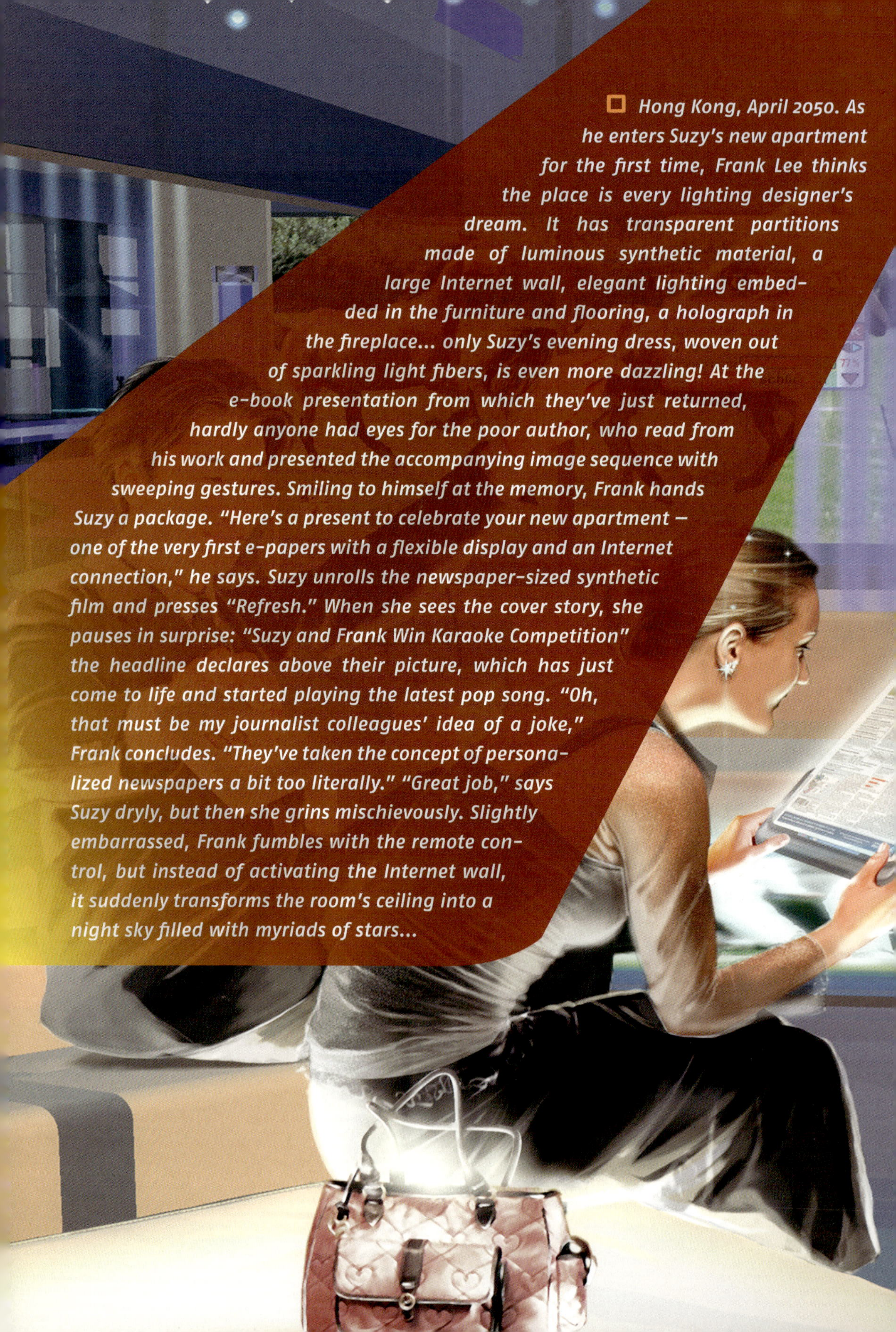

☐ Hong Kong, April 2050. As he enters Suzy's new apartment for the first time, Frank Lee thinks the place is every lighting designer's dream. It has transparent partitions made of luminous synthetic material, a large Internet wall, elegant lighting embedded in the furniture and flooring, a holograph in the fireplace... only Suzy's evening dress, woven out of sparkling light fibers, is even more dazzling! At the e-book presentation from which they've just returned, hardly anyone had eyes for the poor author, who read from his work and presented the accompanying image sequence with sweeping gestures. Smiling to himself at the memory, Frank hands Suzy a package. "Here's a present to celebrate your new apartment — one of the very first e-papers with a flexible display and an Internet connection," he says. Suzy unrolls the newspaper-sized synthetic film and presses "Refresh." When she sees the cover story, she pauses in surprise: "Suzy and Frank Win Karaoke Competition" the headline declares above their picture, which has just come to life and started playing the latest pop song. "Oh, that must be my journalist colleagues' idea of a joke," Frank concludes. "They've taken the concept of personalized newspapers a bit too literally." "Great job," says Suzy dryly, but then she grins mischievously. Slightly embarrassed, Frank fumbles with the remote control, but instead of activating the Internet wall, it suddenly transforms the room's ceiling into a night sky filled with myriads of stars...

A FUTURE WITHOUT BOOKS?

Rarely has an innovation transformed the world as radically as Johannes Gutenberg's invention of the printing press in Mainz, Germany, around the middle of the 15th century. Gutenberg's Bible, Luther's Reformation writings, and the world's first newspaper, which was published in Strasbourg in the summer of 1605, all marked the beginning of a progression towards more self-determination and individual freedom. Nowadays, over 400 million readers around the world buy a printed newspaper every day. But can Gutenberg's invention really survive in an age dominated by television and the Internet? Will we still have printed newspapers and books in 2050?

That may in fact be doubtful. In Europe and the U.S., "the death of the newspaper" is already being discussed, while publishers' sales figures continue plummeting. By contrast, there were already more Internet users than newspaper buyers at the turn of the millennium, and just eight years later their number had almost tripled. On average, Germans spend more time online than they do reading newspapers, magazines, and books combined! About 88 percent of Germans regularly write e-mails, and just as many search the Web for information on travel, tickets, books, and other products. Some 62 percent watch videos and television shows online, either live or at a later time. Worldwide, almost a billion people use the Google search engine, approximately 500 million are members of the Facebook social network, and 350 million consult the Wikipedia online encyclopedia, while more than 220 million people search for products on Amazon and over a billion videos are watched on YouTube every day.

Yet at the same time, news portals also boast millions of users. In fact, the public's interest in news reporting is not declining in favor of shallow entertainment culture; it has merely shifted its focus from printed to digital media. And the advantages of the digital world are very convincing. On the Internet, news items can be updated instantly, and you can watch videos and get more background information with a click of the mouse. And best of all for users, most websites can be visited free of charge thanks to advertising. To date, almost no attempt to get Internet users to pay for news content has been successful. Yet newspaper publishers also stand to benefit from lower printing and distribution costs. Some American newspapers have actually already suspended their printed editions and now appear only online.

Personalized Newspapers and a Book of Books

Some experts think that more and more readers will put together their own online newspaper in the future. One reader might set his personalized news portal to receive updates on important sports events and local news – and the editing system would send him exactly those items every day, or as often as he wishes. Another user, who might be particularly interested in political analysis and in-depth economic coverage, would be able to read a completely different Internet newspaper at exactly the same time. For such a unique service, readers might actually be quite happy to open up their wallets.

According to this scenario, in 2050 a lot of people will subscribe not to printed dailies but to personalized online newspapers, and they will pay to download movies, music or special services such as a navigation system that guides travelers and visitors to places of interest. In Berlin, for instance, such a system might steer you to the Brandenburg Gate and illustrate its history with a few short video clips.

So journalists and editors won't suddenly be out of a job – their research, analytical skills, and ability to sum up complex issues will certainly remain in demand. But the journalists of the future will have to harness multiple media much more than they do today, and let go of the idea of a single, rigid newspaper structure aimed at pleasing everyone's needs. No longer does the Internet simply have to copy the kind of newspaper content that people have been reading for centuries. People's reading habits have clearly changed. Their use of the media has become dramatically faster over the years – just like the speed of film editing. Hardly anyone has the patience to read an entire daily newspaper nowadays. Instead, people take at most a few minutes to get the essentials on the Internet. This trend is being reinforced by the increased use of hand-held devices with Internet access; after all, it's not very enjoyable to read long texts on their miniature displays.

Larger screens are available, though, in the shape of e-books, which are about the size of a small book. In the United States, many best-selling titles are already available in a digital format – often at half the price of a printed book. Approximately a million English titles can be downloaded off the Internet; only a few thousand German titles are available so far. You can also download books while you're on the move. Thanks to the UMTS mobile communications network, it takes less than a minute to download a book onto an e-book. Since it can store up to several thousand books, the device thus becomes "a book of books."

Earning Money Online

In January 2010, Apple announced that in addition to the music, video, and telephone segments, it was going to enter the market for electronic books and newspapers. Steve Jobs then presented his latest triumph: the iPad, an extremely flat PC weighing only 750 grams and shaped like a tablet. The 25-centimeter screen is large enough to display book pages and can be operated intuitively, like all Apple products. Using your finger, you can tap on a book in a small bookshelf and it will open instantly. You can then browse by simply swiping your finger, and enlarge images by spreading two fingers. The iPad is also ideal for watching videos or surfing online.

Many publishers believe that by optimizing the presentation of their products on the iPad and similar devices they can harness the Internet and once again turn a profit with their newspapers and magazines. But they are only slowly beginning to understand what kind of editorial and technical innovations the future may bring. "If you compare online media with the development of the print media, then we're still at the point right after the invention of the printing press," says Rüdiger Ditz, Editor in Chief of Spiegel

Online, one of the German-speaking countries' biggest online portals, with 100 editors and over 110 million page views per month.

The future clearly belongs to the digital media. In 2010 the world's biggest online retailer, Amazon, for the first time sold more electronic books than printed editions. The digital format particularly lends itself to text-books, reference works, and travel guides. Yet even when it comes to fiction, experts are predicting that three quarters of all fiction titles will be sold as e-books by 2020. It's quite possible that by 2050 most households won't actually own bookshelves any more — a few e-books will suffice.

The book of the future won't be static like today's, but will offer a multimedia experience instead. Simply touching the screen will allow you to follow up on cross-references, watch related videos, listen to music that enhances the text, or skip the effort of reading altogether and listen to an audio file of the work at hand. In 2050 people might only keep their favorite printed books (which will have become expensive by then) for nostalgic purposes — and for the pleasure of seeing, touching, and browsing through an actual book. Yet this trend will decline with each new generation of Internet users...until one day printed books may once again become as rare and valuable as Gutenberg's Bible.

Video-Enhanced Newspapers

In the future, newspapers might even take on some of the magic of *Harry Potter.* In J.K. Rowling's novels, the adolescent wizards read The Daily Prophet, a conventional newspaper enhanced by moving images, from which protagonists such as the Weasley family might cheerfully wave at readers. Similar multimedia newspapers might soon become part of our everyday lives — not in the shape of tablet PCs like the iPad, but as newspapers you can fold up. The story behind this technology began in 1990, when the British physicist Richard Friend discovered to his surprise that certain synthetic materials not only conduct electricity but also emit light when placed under a low electric voltage.

Friend subsequently developed the first organic light-emitting diodes (OLEDs), thin layers of plastic less than half a thousandth of a millimeter thick, which produce a green, red, blue or white glow depending on their composition (see p. 109). If you attach many fine electrical conducting paths and switching elements to the back of the layers, you can target specific minuscule areas of the synthetic material — for instance, eight pixels per millimeter – and get them to light up. For a small newspaper format of 26 by 37 centimeters, over six million pixels would be required.

Although it is still extremely difficult to produce an OLED screen with that many pixels at an affordable price and get it to emit light at a steady rate for thousands of hours, researchers all over the world are working hard on the endeavor. Sony has already put small OLED televisions on the market, and Mitsubishi has introduced a huge OLED display measuring almost four meters, which can serve as advertising space – however, it's still only a mosaic of thousands of smaller displays.

The advantages of OLEDs compared to conventional liquid crystal displays (LCDs) are obvious. While LCD screens have a light source behind them, OLEDs glow on their own. They use less electricity, are brighter, offer superior contrast and, what's more, you can read texts on them from the side. They also react to control signals much more quickly, thus avoiding any blurring due to fast movements, such as during soccer games or car races.

These luminous plastics do still have one significant disadvantage. If they're exposed to too much heat or oxygen, or if you leave them in direct sunlight for too long, or if they get wet, they age quickly, lose their brightness or change color. To prevent that from happening, the plastics used for today's displays are hermetically protected by a layer of glass. While that's not a problem when it comes to televisions or computer screens, OLED research scientists still aren't satisfied.

Moving images displayed on plastic film. In the future, organic light-emitting diodes may give us rollable newspapers that can be instantly updated.

Rollable Displays

The "holy grail" they're searching for is a flexible, foldable or rollable display. Using flexible screens, designers could completely transform the cockpit of a car or give advertising pillars a coating to display moving images. They could also easily develop packaging for furniture boxes that shows short videos and animated graphics demonstrating how to assemble the item in question. In the future we'll also be able to transform the pictures on our living room walls into large, flat displays that enable us to surf online, make video calls to friends or family abroad, or download movies off the Internet.

Moreover, the newspaper readers of tomorrow won't have to buy a printed paper to read on the subway or bus. They'll simply carry a thin plastic sheet that can be rolled or folded as required and comes complete with a slim battery good for several hours' use and a wireless Internet connection. At the touch of a button, the sheet would turn into a newspaper that would download the latest articles off the Internet, perhaps even personalized to meet each reader's individual wishes. Just as in the Harry Potter movies, still pictures would suddenly burst into life and show video clips of the best soccer goals, excerpts from the European environment minister's speech at the world climate conference, or images of the first tourists on the moon checking into a brand-new hotel.

If it could be produced at an affordable rate, this flexible, bendable OLED display would undoubtedly open up a new market worth several billion euros, but so far researchers haven't been able to achieve that goal. In principle, there are only two possibilities: either to coat the luminous synthetics with a material that's practically impermeable to water and oxygen while remaining transparent, or to develop OLED materials that are a lot less vulnerable to their surroundings than they are now. The materials expert who manages that feat will not only attain worldwide fame and wealth, but perhaps also the next Nobel Prize in the field of luminous plastics.

That scientist would also deserve a prize for one of the best ideas for reducing waste. After all, the typical newspaper subscriber produces between 100 and 200 kilograms of wastepaper every year, which could be avoided if we used a single plastic film to read electronic newspapers instead. In fact, printed newspapers produce about 50 million tons of wastepaper worldwide every year. Although a large part of that waste is recycled and used to make new papers, the process consumes many billion liters of water and billions of kilowatt-hours of energy. So the development of electronic newspapers would not only offer subscribers a completely new, multimedia-enhanced reading experience, but would also give them the satisfaction of having preserved valuable resources.

☐ Melbourne, May 2050. This is already the tenth patient today with colon cancer — but the procedure has become routine by now. The test to check for a genetic predisposition recommended a heightened level of caution, and a quick blood analysis found proteins produced mainly by cancer cells in the colon. The patient's physician therefore advised a CT scan with a marker that docks exclusively to cancer cells, thereby making them visible. The examination did in fact reveal a small tumor in the colon. Now, a computer automatically guides a catheter tube to an area right next to the tumor, after which the surgeon takes over with a joystick. Thanks to a miniature camera in the catheter, the surgeon can see on the screen in front of him the exact positions where the cancer cells are illuminated in infrared light. The doctor utilizes a tiny chip lab mounted on the catheter tip to analyze another tissue sample just to be sure. With the finding confirmed, he then activates the laser scalpel and uses it to precisely cut the tumor out from the surrounding tissue. After that, he sucks out all of the cancer cells with the catheter. The patient is released from the hospital a short time later, and his doctor assures him that the early detection and removal of the tumor means that he will very probably regain full health. But the physician also advises the patient to consider changing his eating habits...

EARLY ILLNESSES
PET/MR SCAN
MASS SPECTROME
GENOMICS

SOARIAN DSM
HOME PATIENT SEARCH   ADMIN
HISTORY | LAB DATA | IMAGING | GENOMICS | P   ICS | OUTCOME | V   W

Mass Spectrom

490.0
290.0
90.0
4000.0   5000.0   6000.0   7000.0 8000.0   9000.0

Protein expression profiling

SOARIAN DSM
HOME PATIENT SEARCH PREF   ADMIN
TDATA IMAGING | GENOMICS | PROTEOMICS | OUTCOME | OVERVIEW

Colon cancer predisposition gene
Patient: ACGTATTC
Reference: ACGTATTC
SNP-Mutation on Gene HNPCC1
(identified)

Genetic profiling

AGING IN GOOD HEALTH: AN UTOPIAN VISION?

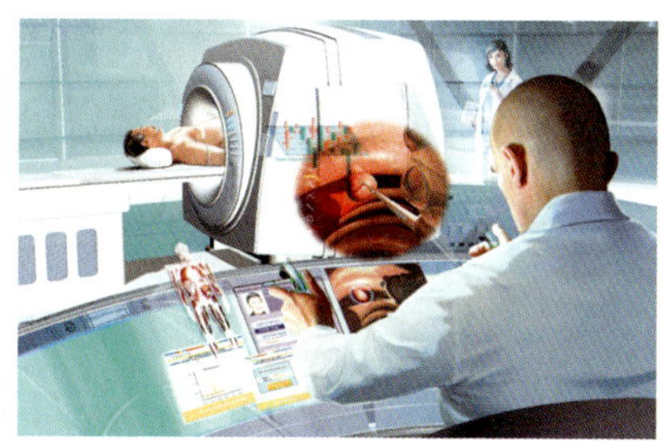

Methuselah still holds the record. He lived to be 969 years old — or at least that's what the Bible says. But since we tend to view such information allegorically, it's likely that the true record for the oldest human being is held by the Frenchwoman Jeanne Calment, who died in 1997 at the age of 122 years and 164 days. Calment, who actually met the painter Vincent van Gogh in Provence when she was a child, is the oldest individual known at the moment whose age at death can be proven with certainty. More and more people are living past one hundred these days, however. At the age of 105, Johannes Heesters performed the role of God in the play *Jedermann* (Everyman), and Oscar Niemeyer, who built elegantly contoured buildings in Brasilia, continued to work in his architectural studio on the Copacabana every day in 2010 — at the age of 102. In the year 2050, people who live to such Biblical ages will be nothing special.

The average life expectancy in industrialized countries today is approximately 80 years — twice as high as it was a century ago. According to the United Nations, there will be more people older than 60 than children under 15 in 2050. This development will be even more dramatic in Germany, whose population is expected to decrease from 82 million today to just 69 million in 2050. One out of every three Germans will then be over 65, and one out of seven will be over 80. This means that for every ten employable individuals there will be 6.4 older people (as compared to 2.7 today), and this will put an enormous strain on the country's retirement system. The impact it will have on the healthcare system will be at least as devastating, because people over 65 account for 40 percent of all healthcare costs. Does this mean we will no longer be able to afford many medical treatments in the future? Or is it conceivable that healthcare might actually become both better and more affordable? Many experts believe the latter is in fact a possibility – if we are able to identify and treat illnesses at an ever-earlier stage of their development. To do this, doctors will have to be able to look deeper inside the human body than ever before – all the way down to the genetic, molecular level.

This applies to many of the most dangerous known diseases. Consider cardiovascular diseases, which account for the highest proportion of the 58 million deaths that occur worldwide each year. Every two seconds, someone dies of a heart attack, stroke, or similar ailment. Whereas cardiovascular conditions account for 30 percent of deaths worldwide, they are responsible for 43 percent of deaths in Germany – or 845,000 people per year.

Cancer occupies second place, accounting for eight million deaths each year around the world, or 14 percent. In Germany, cancer in its hundred different forms is responsible for more than 25 percent of all deaths. Some 25 million people worldwide are now stricken with senility resulting from Alzheimer's disease, and that number could rise to 100 million by 2050.

Such serious illnesses tend to creep up on those they afflict. Sometimes it takes years for the initial symptoms to become noticeable – but metabolic derailment begins long before a patient starts to feel any discomfort; quietly putting his or her body on the wrong track. Scientists around the world are therefore attempting to develop techniques for detecting cancer before tumors and metastases form, and they also want to be able to locate dangerous deposits in the arteries before they block coronary vessels. With cancer, for example, the rule of thumb is that four fifths of treatment costs are generated after dangerous metastases have developed. So if cancer can be identified at an early stage, it can be treated much more cost-effectively and with a greater hope of success. As long as a growth is no bigger than five millimeters, it can generally be treated. Researchers are therefore focusing on devices that deliver increasingly detailed images of the inside of the body, as well as procedures that allow changes at the molecular level to be analyzed. In the future, they will be supported by digital helpers – computers and robots that will act as assisting surgeons.

"Freezing" a Beating Heart

Whereas traditional X-rays only provide a two-dimensional image, the pictures produced by state-of-the-art computed tomography (CT) scans are three-dimensional masterpieces that doctors can use to virtually travel through a patient's body. CT devices have an X-ray source and detectors that race around the patient's body in less than one third of a second. With their resolution of 0.1 to 0.3 millimeters, they make visible the tiniest blood vessels and are even capable of "freezing" a rapidly and irregularly beating heart — and then displaying it in a completely sharp image.

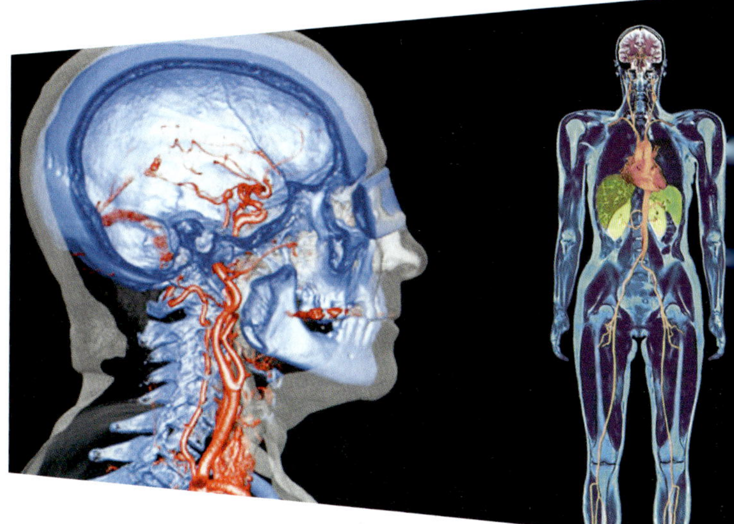

Transparent body. Technologies such as computed tomography and magnetic resonance imaging make even exceptionally small structures visible in 3-D from any angle.

Magnetic resonance tomography (MRT) is also a powerful tool for physicians. MRT devices require no X-rays; instead, they use the signals that certain atomic nuclei (mostly those of hydrogen) send out when they are stimulated by strong magnetic fields. This enables doctors to see soft tissue in the body, such as the brain and other organs, quite well. The most powerful MRT device ever is currently being built in Saclay near Paris. Its 11.7 tesla magnetic field is eight times stronger than that of normal MRT devices used in hospitals and 235,000 times stronger than that of the Earth. Over the next few years, the unit will deliver images of such high resolution that it will be possible to view structures as small as groups of individual nerve cells in the brain.

Scientists at Forschungszentrum Jülich in Germany have linked a 9.4 tesla MRT unit to a PET (positron emission tomography) detector that makes cell activity visible. The examinations carried out here involve injecting a patient with a slightly radioactive substance, usually a sugar, whose radiation is registered by the PET detector. Because tumors need a lot of sugar to grow, the radioactive substance accumulates in their cells, enabling physicians to detect the tumors and their metastases. Brain cells also require a lot of sugar for thinking, which is why they also appear as bright images in PET devices. The new unit in Jülich can not only recognize brain tissue structures with a resolution better than 0.1 millimeters thanks to MRT; it can also utilize the PET detector to make visible brain activity and metabolic processes, which is very important for the early recognition of diseases such as Alzheimer's.

Nanopellets as Medical Messengers

Because MRT is sensitive to magnetic fields, scientists have developed tiny probes that exploit this characteristic. For example, they are using miniature magnetic iron-oxide pellets just a few dozen nanometers in diameter that proteins can be bonded to. These nanoparticles can be made to attach themselves exclusively to inflammatory plaques, thereby making the latter visible in MRT images. The plaques are unstable deposits in blood vessels, and they can easily burst. If they do, they trigger what are called clotting cascades. In such a situation, arteries and veins suddenly become clogged, and this can result in a heart attack or stroke without the patient noticing anything beforehand. Researchers hope the use of nanoparticles and an MRT device will enable the early detection of such plaques in the future and allow physicians to distinguish them from other relatively harmless and stable deposits. Patients thus afflicted could then be treated before anything dangerous happens.

Taken a step further, the nanoparticles could also be furnished with a drug that attacks plaques — or cancer cells, for that matter — directly where they're located. Experiments conducted with animals at Universitätsklinikum Erlangen, Germany, have demonstrated the effectiveness of such an approach. Researchers at the clinic used particles with an iron core ten nanometers in diameter and a starch coating nine times as thick, to which a chemical agent was bonded. The particles were injected into the animals and pulled through blood vessels and toward a tumor with the help of strong magnetic fields. After reaching the tumor, the agent broke away from the particles and began destroying the growth. The nanoparticles themselves were later excreted via the kidneys.

Similar experiments have been conducted with ultrasound devices. Here, scientists use bubbles filled with air or gas (microbubbles) that are clearly recognizable in ultrasound images. Bubbles whose surfaces contain certain molecules can attach themselves to the fine blood vessels in a tumor, or to the protein structures characteristic of circulatory disorders. Microbubbles are well tolerated by the body and can be used to transport medication. Once they reach the tissue that has been targeted, they can be burst using a powerful pulse of ultrasound, after which they release their active agents.

Such procedures with microbubbles and nanoparticles are currently being tested around the world. They will probably be used more and more frequently over the next few decades. It is much less likely that other approaches that are continually discussed as possible future scenarios will be implemented. For example, it is improbable that micrometer or even nanometer-sized technical systems that race through blood vessels and radio their measurement data to the outside world, or actually go into action as a kind of repair brigade will find their way into clinical medicine. Even if you leave aside the fact that very few people will want to have micro-robots inside their bodies, blood is by no means an ideal medium for carrying such devices, especially since it can easily clot.

A Laboratory on a Chip

Micro-technology with its rapidly shrinking components will nevertheless play a major role in applications such as those for labs on chips that can quickly diagnose illnesses. Consider the following example: A runny nose, a cough, a fever, and a feeling of weakness can indicate either a cold or a dangerous flu virus. The only way to be sure is to take a blood sample and analyze it in a special lab — but it can often take days to get the results. In 2050, however, doctors will perform quick tests in their own offices with labs the size of an ATM card.

The basic components of such system have already been developed; these include tiny ducts and pumps that suck in blood drops, integrated chemicals that break open cells, and chambers that capture DNA. These tiny labs will automatically reproduce the DNA they find, mark it with molecules, and transport it to a sensor that will answer what the doctor wants to know — namely, whether pathogens for infectious diseases are present, or whether the patient has certain allergies or hereditary diseases, or is intolerant to specific medications.

Simplified versions of labs on a chip could also be used by patients with certain illnesses to monitor their own treatment progress at home. Just as diabetics can now check their blood sugar levels, patients in the future could use such mini-labs to search their blood for molecules released by cancer cells, and then send the data to their physician, who can determine if the number of such molecules has declined and if a treatment is therefore working. These types of miniature sensors could also be inserted into earlobes like earrings. They could then regularly take blood readings and forward the data via radio.

In 2050, such an ear stud could be used not only to monitor treatments but also for the early detection of diseases — a type of alarm system integrated into the body. That's because not only certain cancer cells but also the cells affected by cardiovascular illnesses form proteins that differ from those in normal body cells. These protein molecules make their way to cell surfaces, and from there to the bloodstream. A mini-lab could pinpoint such biomarkers in the blood and then notify a doctor, who would then attempt to locate their source using a CT or MRT device. Physicians could thus nip ailments in the bud with the help of treatments precisely tailored to the patient in question.

Operating with a Joystick

Physicians of the future will also be able to use mini-labs to treat illnesses in their early stages by placing a lab on the tip of an ultrathin wire or catheter tube. The physician would slide the tube through blood vessels or the colon, for example, until it reaches the tumor, which could then be observed using a tiny camera. The mini-lab's sensors would allow a doctor to determine whether the tumor is benign or malignant. This would be done by injecting a fluorescent substance that docks specifically to cancer cells in the tumor. The cells would activate the substance and cause it to glow. The physician would then be able to see all of the cancerous cells.

Controlling such a catheter is like playing a computer game. The tube has a magnetic tip; the doctor can use a joystick to change an external magnetic field by just a few millimeters in order to guide the catheter precisely to its target. Magnetic guidance makes it possible to navigate even the narrowest curves, such as those in tiny blood vessels. In 2010 a clinic in Nice, France, carried out the first-ever procedure that required no catheter whatsoever. Instead, the patient who underwent this gastroscopy swallowed a small magnetic capsule no larger than a normal pill. The capsule contained a miniature camera that could be precisely controlled by a magnetic field.

Magnetic guidance could also be performed automatically in the future. Here, doctors would simply mark the target location in precise 3-D images produced by a computed tomography device, and the computer would calculate the best path for the catheter and then guide it there. Surgeons in the year 2050 will simply sit in front of a large viewing device and monitor the catheter's journey.

They will see each millimeter of the body as a giant image and will be able to use special gloves to remotely control the tiny instruments contained in the catheter tip. They will even be able to receive feedback via their sense of touch, such as a sense of how much pressure micro-scissors or touch sensors exert upon a small tumor in the body. Medical visionaries refer to this as "immersion surgery."

In the future, surgeons will also increasingly use augmented reality techniques to perform operations. Such techniques are currently being tested by Nassir Navab, a professor at the Technical University of Munich. The idea is that a surgeon would start out with CT or MRT images and, with the help of a 3-D head-mounted display, superimpose these images over the actual target area to be operated on, thus making the patient appear transparent. "When a trauma surgeon places a scalpel on a patient's skin in the future, he or she will see the bones and possible breaks underneath, so it will be clear where the incision has to be made," says Navab.

Three Billion Letters in the Book of Life

The most ambitious vision for the year 2050 is that of personalized medicine — diagnoses and treatments perfectly tailored to the individual patient in question. But in order to understand why some people fall victim to some illnesses and others do not, or why a certain medication works for one patient and not for another, doctors need to be able to look deeply into the interior of cells and study their genetic makeup.

The human genome — DNA — is made up of three billion letters (base pairs) that together form around 30,000 genes. Each gene in turn contains either a blueprint of proteins or commands that stimulate or stop the synthesis of certain molecules. If a cancer researcher wants to find out what distinguishes a tumor cell from a normal one, he or she will look for differences in the DNA, genes, proteins, or molecules on the cell surface. This is an extremely complex task, because the many thousands of molecules in a cell are constantly interacting and these relationships also change over time.

A milestone for future medicine was achieved when the entire human genome was decoded at the beginning of the 21st century. Among other things, this accomplishment makes it possible to trace hereditary diseases back to their genetic origins and find the mutations that make individuals immune to the pathogens that cause diseases such as AIDS or malaria. Still, knowing the sequence of the base pairs in the DNA is not enough in most cases, because we also need to determine which genes are different in healthy and sick people. Some 1,500 genes that are altered in afflicted cells have been discovered to date — but even that usually won't be enough, since scientists also need to know which proteins are active or inactive at a certain point in time.

Proteins are the tiny machines that keep cells alive by transporting molecules, triggering or stopping chemical reactions, and recognizing signaling substances (semiochemicals). Even when genes are identical, the creatures that carry them can still look completely different due to variations in the activity of the proteins that are also present. For example, a caterpillar, a pupa, and the butterfly it becomes all have the same genes, as do a tadpole and a frog.

All human cells also contain the same genes, but these perform different functions depending on whether they end up becoming part of skin or muscle cells, a nerve cell, or a sperm cell. The number of proteins present in human cells ranges from several hundred thousand to several million, which shows the immense size of the challenge facing scientists who seek to understand cell processes.

From 25 Years to 14 Days

Automated laboratory technology has made cracking the genetic code a routine task these days. Whereas it took 25 years to decode the nearly 10,000 base pairs in the HIV virus that causes AIDS, scientists were able to crack the SARS virus, which leads to pneumonia, in just a few weeks. And they only needed 14 days to map the H1N1 virus that causes swine flu. Just two days after its genetic code was published, researchers used high-powered networked computers to identify the locations in the DNA of the H1N1 type they were studying that are completely different from those in any other type. Such precision made it possible to develop a test for the flu virus in a very short time.

Human genome decoding has now also been automated, leading to a dramatic decline in costs. The Human Genome Project took 13 years to carry out and it had a price tag of around $3 billion, but the man who discovered DNA structure, James Watson, was able to have his own genome decoded in 2007 for only $1 million.

Several companies now actually offer complete DNA decoding within four weeks for the price of a mid-range sedan. It will therefore probably take only a few years until each individual can obtain a copy of his or her own genetic code for $1,000 and bring it home on a USB stick — in the hope that it doesn't harbor any incurable hereditary diseases.

Researchers can now use computer simulations to activate and shut down genes and then observe the effects. They're doing this not just in order to isolate cancer genes but also to revolutionize pharmaceutical research. For example, it's already known that genetic differences are the reason why certain medications have no effect on some people, as well as different side effects on different patients. If scientists knew which genes were responsible, they could predict the effect medications will have on certain individuals and make it easier for pharmaceutical researchers to achieve their goal of producing drugs tailored to individuals so as to achieve the greatest possible healing effect and the fewest possible side effects — a kind of "personal pill."

Is DNA fate? A person's genetic makeup determines the course of his or her life from birth to death — but scientists are starting to rewrite this operating manual.

Digital Assistants

The healthcare systems of tomorrow will require high-powered computers not just for the aforementioned simulations but also to manage the flood of images produced by state-of-the-art medical equipment. Ten years ago, a computed tomography examination generated about 50 images — today the same examination produces over 2,500 pictures, far more than a single physician can analyze. That's why the first intelligent image evaluation procedures have been developed. These can comb through hundreds of tomography images in just minutes and then call the radiologist's attention to those that perhaps show a small breast or lung tumor only millimeters in diameter, or a polyp in the colon.

These systems have learned from examples how to identify the characteristics of such tumors, and this enables them to detect similar growths in other images. This is the only way to ensure that doctors who have to make dozens of diagnoses a day won't miss a suspicious growth. Such systems will also increasingly be equipped with voice command features in the future. Physicians will then save a lot of time, as they will be able to issue spoken orders such as "Show me the lower left pulmonary lobe and compare it with the image from the last examination."

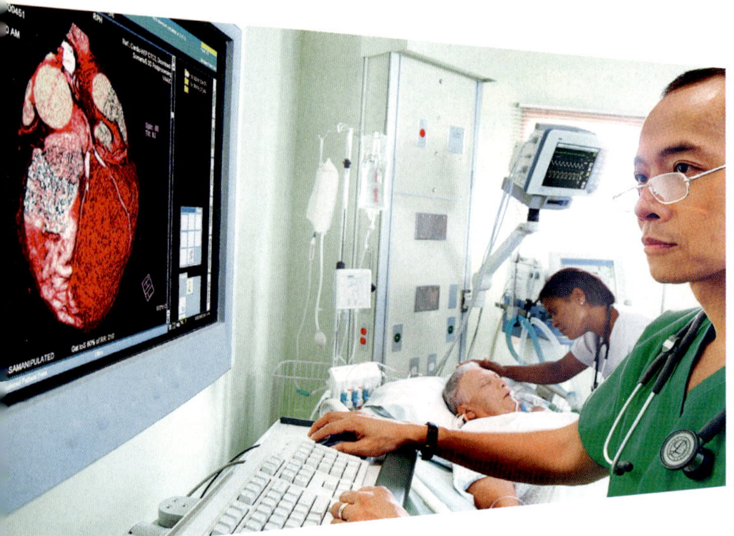

Computer assistance. A computed tomography examination today produces more than 2,500 images. Learning-capable software can alert doctors to abnormalities in this flood of data.

Also now in development are digital assistants that can link CT and MRT images with lab analysis results and patient-related sociological data. Such assistants will eventually by able to produce recommendations for the most promising treatments. Particularly helpful here are studies that examine the connections between illnesses, lifestyle factors, and genetic predisposition. One of the most extensive such studies anywhere is being conducted in Mecklenburg–West Pomerania in Germany, where thousands of residents have been thoroughly examined for years. They've been given full-body scans with MRT devices, have undergone examinations of their hearts, eyes, and teeth, and have provided blood and urine samples.

Researchers are studying their DNA and sleeping patterns, and staff from the University of Greifswald have been extensively questioning them about their eating habits, whether they smoke or drink, and what medications they take. The scientists' intention here is to clarify many questions such as: Do lifestyle and living conditions play a role when women develop breast cancer? Is there such a thing as a genetic predisposition to liver and kidney disease? Why are gallstones significantly more common in Mecklenburg–West Pomerania than in the world as a whole, with the exception of one group of indigenous people in Chile? Can poor teeth in childhood affect growth, or even increase the risk of a heart attack in adults? The ultimate objective of this research is to identify new approaches to treatment and tailor them to specific patients.

A Second Opinion for Timbuktu

Advances in information and communication technology will also ensure smooth data exchange between hospitals, doctors' offices, pharmacies, and insurance companies. Among other things, the future will see the still largely paper-based patient administration system replaced by electronic patient files, while traditional X-rays will increasingly give way to digital images. This will help to avoid redundant examinations. This digital medical world is already a reality at the Rhön–Klinikum AG hospital group, whose patients (more than one million each year) are set to benefit from electronic patient files that contain digital examination images, lab results, and medical history profiles. Physicians who need to obtain information quickly in hospital wards and examination or operating rooms can simply look at the data. Of course, access to this data is highly restricted. Patients must approve access to their data, which can then only be viewed by the doctors who are treating them. A record is also kept of each time patient data is called up. If necessary, external specialists can be called in to discuss specific cases online, for example, in "tumor conferences." In the future it will be common to quickly obtain a second opinion from an expert in this manner — regardless of where the patient or specialist is located. In other words, it will be just as easy for a radiologist in Nepal or Timbuktu to consult a colleague in Chicago and show him or her a CT or MRT image as it would be for a doctor in Detroit to do the same thing.

And because more and more operations in the future will be robot-assisted, it's even conceivable that a surgeon in London might guide a robot even as the patient lies on an operating table in the African jungle. Many doctors are against such telemedicine applications, however. They claim that something can go wrong in any operation, in which case the responsible surgeon would be thousands of kilometers away. They also insist there should always be a well-trained medical team on site — but of course that would make a remote operation unnecessary in the first place.

Wearable Sensors and Motivating Fish

One objective advanced technology can surely help to achieve is the goal of enabling the elderly or chronically ill to live a relatively fulfilling life within their own homes for as long as possible. This will be done through the use of intelligent processes for self-diagnosis and remote medical monitoring. Sensor systems already exist that measure weight, blood sugar, pulse, and blood pressure and send the results to a primary physician or nurses, who can then contact a patient if his or her condition appears to be deteriorating.

In the future there will also be more and more sensors that can be worn, as they will become so small that they can be easily integrated into clothing. The "Smart Senior" project in Berlin, for instance, is currently testing a wristwatch equipped with a sensor that is fed data on typical movement patterns such as walking, climbing stairs, or lying down, and then compares this information with current values. The device therefore knows if a senior has fainted — for example, it will notice the absence of the usual micro-movements made by his or her arms during sleep. It's also possible for a senior to wear a special Band-Aid on the upper arm that measures his or her pulse, body temperature, and even blood oxygen content. It transmits this data to the watch, which then forwards it via WLAN to a mini-computer in the apartment. The latter is in turn connected via the Internet to the Telemedicine Center at Charité Hospital. Doctors are notified in the event of an emergency — and seniors don't even have to remember they're wearing the devices, which function fully automatically.

Another monitoring system, this one rather amusing, is known as "Fish 'n' Steps." American researchers have used it with obese people to motivate them to pursue a more active lifestyle. The idea is to get overweight individuals to take at least 10,000 steps per day. Participants in the study were given a step counter and their computers were equipped with a virtual aquarium in which fish swam. When a test subject took more steps than the daily minimum, his or her fish grew; if he or she took fewer steps, the fish began to waste away. The fish got the participants' pride going. But even more interesting was the peer pressure that developed. Each aquarium housed fish belonging to four participants. If one of the fish failed to grow for a week, the water in the aquarium became murky. The lazy person who was responsible for this state of affairs then started moving more because he or she didn't want to look bad in front of the others.

This example indicates that the healthcare systems of the future will use increasingly sophisticated methods to detect and treat illnesses at the earliest possible stages. It will also be just as important to make sure people don't get sick in the first place — in other words, to help them retain a good level of health rather than getting them to regain it. One example of this can already been seen in the form of anti-smoking attitudes around the world, which have increased in part because of the knowledge that smoking is a major cause of lung cancer and cardiovascular disease. According to the World Health Organization, cigarettes are responsible for nearly one out of every ten deaths worldwide. In Germany alone, more than 110,000 people die each year from smoking-related illnesses, with an additional 70,000 or so deaths caused by alcohol abuse, 1,300 by illegal drugs, and 4,000 by traffic fatalities. None of these deaths are an unavoidable fate.

Other causes of death can also be prevented. Particularly disturbing is the fact that nearly 12 million children under the age of 15 die every year worldwide. The causes here, along with complications during birth, often include diseases such as cholera, typhus, and dysentery (2.1 million deaths per year, mostly children), respiratory infections such as pneumonia (nearly four million deaths, also mostly children), and malaria (one million deaths). Many of these diseases could be prevented through better hygiene and clean drinking water, or could be treated with relatively inexpensive drugs. The same is true of cardiovascular disease and diabetes in the industrialized nations, where a healthier lifestyle, regular exercise, and more moderate consumption of alcohol and tobacco products would also prevent many illnesses.

Munich, June 2050. Practice makes perfect — even for surgeons. In these advanced times it has become standard procedure for physicians around the world to reserve a couple of hours in their clinic's virtual reality operating room (VROR) and, as part of a team, rehearse the treatment of complicated cases in real time. Equipped with 3-D goggles and skin-tight haptic gloves, surgeons simply download a virtual but highly lifelike patient onto the operating table and complete a dry run of different kinds of surgery. This might involve maneuvering artificial heart valves into place with a minuscule catheter or replacing tissue damaged in a heart attack with reprogrammed stem cells — or, in an even more complex procedure, implanting microchips made of millions of photodiodes under the retina and then connecting them to nerve cells in order to restore the sight of a blind patient. Even an operation of many hours, such as the attachment of a prosthetic hand, can be rehearsed in the virtual reality operating room. In this case, the robotic hand is connected via nanoscale filaments to nerve fibers in the patient's arm. If all goes according to plan, the patient will be able to move the new hand at will and experience sensation right to the very fingertips. Even with the computer and robotic support available in 2050, such operations will be the preserve of only the very best surgeons. And even they will need to practice beforehand in the clinic's VROR.

TODAY'S MEDICAL MIRACLES: TOMORROW'S ROUTINE CARE

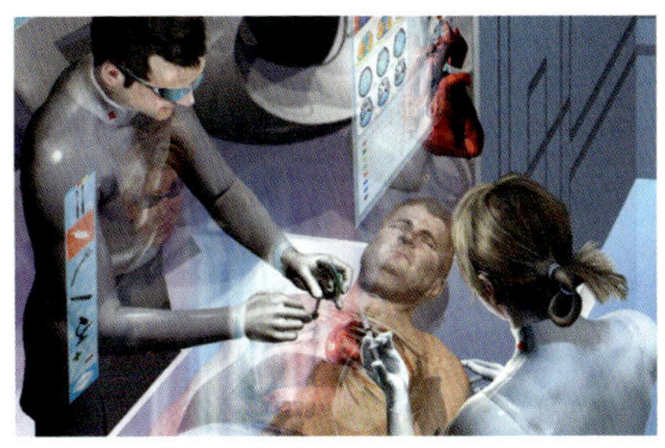

By the year 2050, what might now look like miracles of medical science will have lost all of their capacity to surprise. Many procedures that are now being researched but that still seem like science fiction today – operations to restore sight to the blind, hearing to the deaf, movement to the paralyzed, and even limbs and robotic hands to victims of accidents – will become standard medical procedure in coming decades. There is no such thing as "never," particularly in the field of medicine, where seemingly incontrovertible truths can be toppled from one day to the next.

Examples of this are numerous. Take Barry Marshall, an Australian physician. He and his colleague Robin Warren turned medical orthodoxy on its head in the early 1980s with their assertion that peptic ulcers were not principally the result of too much stomach acid, itself associated with stress. Instead, they were confident that they had identified the real culprit in the shape of a bacterium named Helicobacter pylori. Medical opinion was highly skeptical at first, not least because it was thought that no bacteria would be able to survive in the acid environment of the stomach.

Marshall's reaction was to infect himself with the bacterium as a way of proving the theory. He therefore mixed himself a cocktail of Helicobacter pylori and swallowed the dreadful concoction in one gulp. The results were exactly as he had expected. He developed powerful stomach pains and was barely able to eat. "On the eighth day," he recalls, "I woke up and ran to the bathroom to vomit. The mass of bacteria in my stomach had removed all the gastric acid." Fortunately, he had already developed a remedy and was able to cure himself by administering a bismuth-based antibiotic. Even after such an ordeal, it was still a number of years before skeptics fell silent. Finally, in 2005, Marshall und Warren, two unconventional researchers, were awarded the Nobel Prize in Medicine.

Microchips for Eyes and Ears

Restoring sight to the blind would represent a similar medical sensation. That is exactly what Eberhart Zrenner, Director of the Institute for Ophthalmic Research at the University of Tübingen, achieved in December 2009. At a congress in Miami, Zrenner described a four-hour operation during which a microchip measuring 3 x 3 millimeters was implanted under the retina of Miika T., who had been without sight for over 20 years.

"For the first time ever, a patient fitted with a visual prosthesis has crossed the boundary beyond which he is no longer regarded as legally blind," he said proudly. The chip contains 1,500 photodiodes that convert incident light into electrical impulses and transmit these signals to those nerve cells in the retina that are still intact. The rods and cones in the patient's eye, which normally undertake this task, had been destroyed by a genetic disorder known as retinitis pigmentosa, which affects three million people worldwide.

When the chip was switched on, the patient was able to see for the first time in over two decades. Initially his vision was very unsteady because his brain first had to become re-accustomed to the signals from the optic nerve. Within a few hours, however, he began to recognize objects and was able to distinguish an apple from a banana and even detect a spelling mistake in his own name, although when written in letters several centimeters high. In other words, the operation sufficed to help him find his way around and see a rough but recognizable image of people.

In the future it should be possible to pack even more photodiodes onto a microchip. This means that many of Miika's fellow sufferers have good reason to hope that they will one day regain their sight. In the future, retinal implants will become as common as cochlea implants are today. The latter comprise a microphone and a speech processor, among other hardware, and are suitable for patients who have lost the functionality of the hair cells in the inner ear but still have an intact auditory nerve. Learning to hear again after an implant has been fitted can be as complicated as learning a foreign language, but sooner or later patients do regain the ability to understand normal speech.

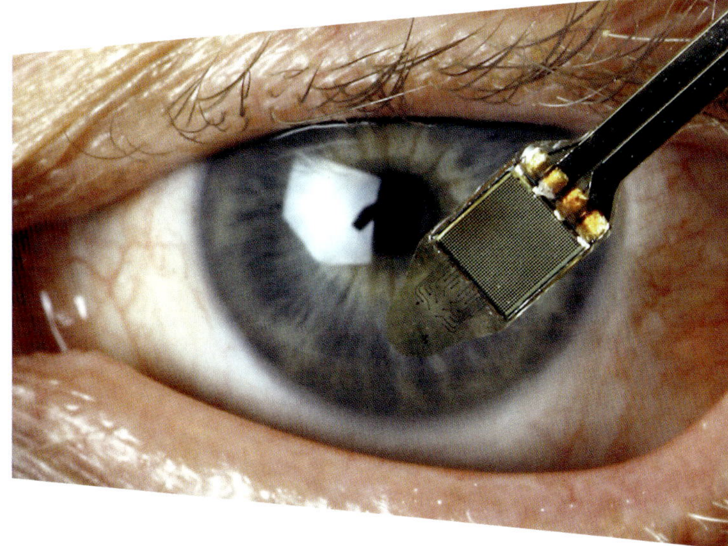

Subretinal implant. Using a microchip with 1,500 photodiodes, doctors have succeeded in restoring sight to a blind person.

Mind over Aluminum Fingers

Alongside these advances in implant technology for the eye and ear, physicians have also succeeded in replacing, with an electronic prosthesis, that most complex of prehensile tools: the human hand. In November 2008 a team of Italian surgeons, neurologists, and bioengineers attached a robotic hand to the arm of 27-year-old patient Pierpaolo Petruzziello, who had lost his hand in an automobile accident. The new hand, which comprises five aluminum fingers (including a thumb), was connected to nerves in the patient's arm via four electrodes and fibers a mere ten nanometers thick. Following the operation, Petruzziello was able to control his new hand just by the power of thought. Not only could he grip objects with the hand, but he also regained a sense of touch. The prosthetic hand will now undergo a trial phase of several years, before being permanently transplanted. The message behind such medical breakthroughs is clear: Tomorrow's world will be increasingly populated by cyborgs — humans with bodies that permanently include electronic components.

In a further advance, physicians will also be able to perform bodily repairs on the genetic and cellular level. In addition to organ transplants between humans or between animals and humans, there will also be stem cell transplants to treat conditions such as leukemia, whereby stem cells from donor bone marrow become active in the patient's body. Researchers are also hopeful of being able to reprogram stem cells, and thus transform them into other types of cell, including heart cells and neurons for the replacement of damaged cardiac and nerve tissue.

Even without stem cells, it is already possible to cultivate certain kinds of tissue such as skin, cartilage, and bone. To this end, a tissue sample is removed from the patient and then grown in the lab, before being reintroduced into the body. The virtue of this procedure is that the immune system does not reject such tissue, because it is recognized as coming from the patient's own body.

The use of gene therapy will also increase. This involves the introduction of properly functioning genes into cells containing defective genes. As a rule, the so-called vector — a vehicle used to transfer genetic material to a target cell — is a virus, which penetrates the cell and deposits its "cargo" there. This method is both difficult and highly controversial because of potential adverse effects such as acute inflammation. To date, there have been some 1,500 clinical studies in the field of gene therapy. Of these, there have been quite a number of successes, including the use of gene therapy with stem cells and white blood cells.

Combating Malaria with Sugar Molecules

Peter H. Seeberger, Director of the Max Planck Institute of Colloids and Interfaces in Potsdam, is currently investigating the sugar molecules that are found on the surface of the cells of parasites and bacteria. He and his team have built a system that functions in a manner similar to that of a fully automated DNA or protein synthesizer, but with the difference that it will produce carbohydrates from sugar molecules as required. This carbohydrate synthesizer may someday become as important as DNA and protein synthesizers are today, because it facilitates the rapid development of vaccines.

Specific carbohydrates located on the shells of pathogens are the best point of attack for immune cells. This makes them ideal for preparing the immune system to counter microbial invasions. "Vaccines train the immune system to recognize certain molecules on the surface of an infectious organism and react more quickly," Seeberger explains. "With our synthesizer, we've been able to produce not merely one but a large number of carbohydrate structures." In turn, this provides researchers with a highly efficient means of testing various carbohydrates to determine their suitability for use as vaccines against viruses, bacteria, or parasites.

As it happens, Seeberger's team did discover a carbohydrate on the shell of the malaria pathogen by means of which the parasite gains access to the red blood cells of the human body. With the help of the carbohydrate synthesizer, the group was then able to develop a vaccine that is now heading for clinical trials. "To my knowledge, ours is the first vaccine that might be able to prevent malaria," says Seeberger. This is because the vaccine blocks the sugar structures on the surface of the pathogen, with the result that it can no longer penetrate blood cells and make people ill.

Researchers hope to develop a vaccine, in combination with proteins, that would be particularly suitable for children under the age of two, the patients who suffer most from malaria. If the scientists succeed, it may well represent a decisive victory in the battle against one of the biggest scourges of mankind. There are currently 2.5 billion people at risk of contracting malaria worldwide, and 300 million already carry the disease. The genome of the malaria pathogen has been identified even in the tissue of most Egyptian mummies that have been examined. Many researchers assume that Tutankhamun died of the disease. Yet the fight against malaria has been bedeviled by setbacks. A vaccine that worked well in South America, for example, was largely ineffective in Africa.

Trials are now being conducted on 20 to 30 promising candidates for vaccines based on a variety of approaches. The most advanced tests are those based on a substance containing proteins from the early stage of the life cycle of the parasite that is responsible for causing malaria. The vaccine is currently being tested on a sample of around 16,000 children in seven African countries.

This kind of development is being spearheaded by companies including GlaxoSmithKline and institutions such as the Malaria Vaccine Initiative, which has received funding of over $200 million from the Bill & Melinda Gates Foundation. Bill Gates, the founder of Microsoft, chooses a drastic comparison to explain his commitment to this cause: "More money is invested into drugs for baldness than into malaria."

Reversing Paraplegia

Speaking at the Falling Walls Conference of November 2009 in Berlin, Martin Schwab, Professor of Neuroscience at the University and Swiss Federal Institute of Technology in Zurich, discussed another remarkable breakthrough in medical science. "For 100 years," he said, "we have taught medical students that injured nerve fibers in the spinal cord and brain do not re-grow." Researchers have tried injecting growth factors, but to no avail. In 1989, the year the Berlin Wall fell, Schwab realized why. "There is no point in trying to stimulate growth in injured nerves. Instead, the inhibitory factors that prevent growth have to be blocked," he said. With this assertion, he flew in the face of accepted medical opinion, which had refused to countenance the existence of "growth inhibitors" in the nervous system.

Schwab was able, however, to identify not merely one such protein, but an entire family of them. He named the most important of these "Nogo-A." This tells the nerve fibers: "Stop! No further growth!" However, if this protein is neutralized by suitable antibodies, nerves begin to grow again – and also do so along and around the site of an injury. The result is that paraplegia in laboratory animals can be reversed. As Schwab explains, rats that have had nerves in the spinal cord severed "are able, after three to four weeks of treatment, to negotiate narrow wooden beams. In a similar experiment, macaque monkeys regained the use of their hands."

Such experiments are among the few that necessitate the use of living animals. Whereas many drugs can be computer-simulated and then tested on human cell and tissue cultures, studies examining injured nerve fibers and their possible re-growth have to be carried out on living organisms in order to demonstrate the efficacy of the procedure under investigation.

In cooperation with Novartis and a network of clinics in Europe and Canada, Schwab's team is now carrying out tests on patients. In particular, researchers hope to be able to help people whose injury has occurred less than two weeks before and who do not have any other serious wounds. It is not a question of patients being able to go dancing again, or playing soccer or the piano, researchers say, but of achieving an improvement in general mobility. That would be an enormous gain. The results so far seem encouraging, with a sample of 40 patients showing high tolerance to the antibodies and no adverse effects. Should the agent have the desired effect in a second phase, involving tests on a further 150 patients, this would represent a huge breakthrough in the treatment of paraplegia.

It seems barely plausible that future paraplegic patients might merely have to swallow a pill or receive an injection in order to regain movement of their hands and legs, as though nothing had happened. Yet what this example shows is that in the field of medicine, as in many other areas of science, nothing should ever be considered impossible. It is always worthwhile to question received opinion and to try a new approach to a well-known problem. Whether we mean Albert Einstein's theory of relativity, the quantum physics of Max Planck, Muhammad Yunus' development of the concept of microcredits, Barry Marshall's ulcer-causing bacteria, or Martin Schwab's nerve growth inhibitors — without such people who dare to think outside the box and relentlessly pursue an idea, many key breakthroughs in science, politics, and technology would never have come about.

AFTER 2050 – BEYOND THE CRYSTAL BALL

If all indications are correct, in the first half of the 21st century the human race will have to deal with one overriding challenge: the restructuring of its entire energy system. It will have to move away from fossil resources such as coal and oil and toward a culture of sustainability with a high proportion of renewable energies, such as solar and wind power, geothermal energy and biomass, and water and wave power. This revolution is necessary for two reasons: first, because of the depletion of our fossil resources; and second, because of the massive changes to our environment that will result if we continue to plunder nature unchecked.

If the human race manages to achieve this restructuring peacefully, it will make a huge step toward safeguarding a permanently livable world. But this also implies the simultaneous achievement of the United Nations' development goals – overcoming poverty, inadequate education, starvation, lack of drinking water, environmental pollution, and infant mortality. Otherwise, terrorists and warmongers will receive a steady stream of new recruits, and they will exploit the conflicts that accompany every restructuring process for their own ends. Instead of putting a sustainable system for using energy and raw materials in place, the world might then collapse in chaos.

This emphasis on sustainability is an essential aspect of the sixth Kondratiev cycle that is described in the first chapter – a cycle that is just beginning and will characterize the period ending in 2050. The focus of this cycle will be on health in a holistic sense – the health of the environment and the health of human beings. This second aspect includes the struggle against infectious diseases such as malaria and AIDS as well as typical causes of death in the industrialized countries, such as heart attacks and strokes, cancer, and in particular the illnesses accompanying advancing age, such as eye diseases, diabetes, and Alzheimer's disease.

The focus on human well-being could be the key element that defines the second half of the 21st century — and it would go far beyond the struggle against diseases and the extension of human longevity. In addition to the goal of ensuring a healthy old age, the focus would be on the essential core of a human being — the brain and all the other complex processes taking place in our bodies. Understanding what goes on inside these tiny gray cells and all the other cells of our bodies, as well as our immune systems will probably be one of the great scientific challenges for decades to come.

And it will become ever more mysterious and wonderful as we investigate smaller and thus more fundamental details. How do thinking, learning, and feeling actually work? What controls our behavior, and what exactly is consciousness? And do these concepts apply only to human beings, or will it be possible to create manmade systems that understand and have self consciousness? These questions will certainly occupy researchers throughout the 21st century.

The Basic Elements of Life

Tomorrow's researchers will penetrate ever deeper in their investigations of the basic elements of life and attempt to understand how they interact. For example, they will examine how nerve cells in the brain interact, with the help of biochemical messenger substances and electrical signals, to produce thought processes. They will study how the network of genetic material, proteins, sugar molecules, and other substances functions within each individual cell, as well as the essential aspects of the aging process. And they may even be able to stop some aging processes — but possibly only at the cost of a higher risk of cancer.

At the same time, synthetic biology will gain tremendously in importance. Researchers already have thousands of "biobricks" in their toolboxes. These biological components are fragments of genetic material that carry out certain tasks and can be combined with one another almost at will. Scientists can smuggle them into cells, and in the future they will even be able to design and produce synthetic cells on the drawing board. The aim is to eventually use such biofactories to produce medications and biofuels in a very precise and efficient way.

Whereas biologists are focusing on genes, proteins, and nerve cells, computer specialists are dealing with the bits that are being processed ever faster in ever larger quantities, and materials scientists are investigating atoms and photons. Here too, the aim of future research will be to make these tiniest elements manageable. Researchers have already demonstrated that it is possible to manipulate atoms and photons, use them for calculations, and build things with them. Photons can be used to create new communication and computer systems — for example, systems that are basically impossible to eavesdrop on.

Individual atoms can also be used to build components with nanometer dimensions, such as the "quantum points" that can be connected to form larger computational units, or components that are so flat that they consist of only a single layer of carbon atoms, forming the material known as graphene. This substance, for which the Nobel Prize in Physics was awarded in 2010, is transparent, extremely compact, and an excellent conductor of electricity. Researchers hope they will one day use it to create extremely thin computer touchscreens, innovative solar cells and very fast switching elements.

But the flurry of hype that arose a few years ago with regard to nanotechnology may have been exaggerated. Nanorobots acting as "health policemen" inside our bodies belong in the realm of fantasy, as do nanocomputers implanted in our brains to serve as auxiliary storage areas or mobile radio interfaces. Equally fantastic are "nanoassemblers" that would put together products of all kinds from atomic components at will, just as children combine Lego pieces. Apart from the fact that many of these ideas are unacceptable on ethical grounds, they bump up against very practical and physical limits. Besides, nanoassemblers don't have to be invented — they already exist. The ribosomes in cells are perfect nanoassemblers. They put together proteins according to the instructions that are stored in DNA.

The nanotechnology of the future will therefore mainly be applied to projects of a completely different kind, just as it is today — for example, developing new catalysts for speeding up chemical processes or using carbon to build nanotubes that have thousands of times the electrical conductivity of copper wires with the same dimensions. Researchers will also cover materials with extremely thin coatings in order to protect them from high temperatures, make them dirt-resistant or change their degree of transparency. Today, materials whose structure makes them invisible to microwaves already exist. With the help of nanotechnology, a "cap of invisibility" may one day be developed for optical light as well.

Destination: A Small Blue Planet

As a result of all this, scientists won't be running out of exciting fields of research anytime soon. Even a permanent base on Mars that the futurologists at the RAND Corporation expected to see by 2010 will exist one day. The only question is when. But during the coming decades our money will certainly be better invested on Earth. We still have a number of challenges to meet here at home before the human race can embark on the conquest of space. Astrophysicist Carl Sagan once wrote the following thought after seeing a photo of the small blue planet Earth that astronauts had made from the moon: "In our obscurity, in all this vastness, there is no hint that help will come from elsewhere to save us from ourselves." And when Neil Armstrong was asked whether he didn't feel like a giant when he stood on the moon and was able to blot out the distant Earth with his thumb, he answered, "I didn't feel like a giant. I felt very, very small."

Our first priority is therefore to preserve the Earth as a habitat for ourselves and for the other creatures that live here with us. That is the challenge of this century, and tackling it will be worth our while. This will not be possible without new ideas and inventiveness, but society must decide which technologies it wants to use, and it must then do its utmost to make sure

Spaceship Earth. Before the human race can return to the moon and visit other planets, it has some challenges to meet on Earth in order to make sure our blue planet remains a home worth living in.

they are used sensibly. These discussions must not be postponed, and these decisions must be made, because even though every course of action harbors risks, so does every failure to act. In view of the tremendous challenges the human race is facing today, doing nothing is not an option — and it would be unworthy of us as human beings.

Numerically speaking, the overwhelming majority of all the scientists and innovators who have ever lived are alive on our planet today. The range of possible activities and research areas was never as broad or as vital as it is today. Finding the right ideas that will change the world is a goal worth pursuing, and it's fun to generate ideas that are new and sometimes outside the box. After all, this thought expressed by Richard Smalley, the discoverer of a new form of carbon, is still valid: "When a scientist says something is possible, they're probably underestimating how long it will take. But if they say it's impossible, they're probably wrong." Joining in the work of creating a world worth living in is certainly not a "mission impossible."

Many of the insights and trend statements mentioned in this book are the result of ten years of work for the international research and innovation magazine *Pictures of the Future* (www.siemens.com/pof), of which I am the Editor-in-Chief and for which Arthur F. Pease is the Executive Editor, English Edition. I would therefore like to thank all of my colleagues and the many freelance authors who have worked for this magazine. In particular, I would like to thank Dr. Norbert Aschenbrenner, Arthur F. Pease (also for editing the English edition of this book), Florian Martini, Sebastian Webel, Ulrike Zechbauer, Dr. Andreas Kleinschmidt, the layout team headed by Rolf Seufferle, including Rigo Ratschke, and Jochen Haller, the translation team of Transform, Cologne, and Judith Egelhof's picture editing team. Many of the illustrations featured at the beginning of the various chapters were created by Natascha Römer and Maxim Osadtschij (www.roemer-osadtschij.de). I am particularly grateful to them for providing us with these imaginative renderings of the future.

In addition to conducting my own research, I obtained information from articles in newspapers and magazines such as *Süddeutsche Zeitung*, *Frankfurter Allgemeine Zeitung*, *Berliner Zeitung*, *Die Welt*, *Die ZEIT*, *Bild der Wissenschaft*, *Scientific American*, *Technology Review*, *Focus*, and *DER SPIEGEL*, as well as from numerous websites, particularly those belonging to the German edition of Wikipedia. These sources are a rich mine of information, whereby it is always advisable to consult one or two additional sources to ensure the data is correct. The presentations at the Falling Walls Conference in November 2009 were another valuable source of information. The conference, which was held to commemorate the 20th anniversary of the fall of the Berlin Wall, was attended by leading researchers from all over the world, who described a variety of breakthroughs and visions from a wide range of areas.

In spite of all the careful research, it is not possible to assume liability for the accurateness of the content or all the quotes. Due to the compact nature of the presentation, simplification was sometimes necessary. I hope you will excuse any misinterpretations that may have arisen as a result.

Finally, I would particularly like to thank my wife Angelika, my children, my mother, and my friends for showing so much patience during the many weekends and vacation days that I had to sacrifice in order to research and picture life in the year 2050.

PICTURE CREDITS